CONTROL OF SPATIALLY STRUCTURED RANDOM PROCESSES AND RANDOM FIELDS WITH APPLICATIONS

Nonconvex Optimization and Its Applications

VOLUME 86

CONTROL OF SPATIALLY STRUCTURED RANDOM PROCESSES AND RANDOM FIELDS WITH APPLICATIONS

By

RUSLAN K. CHORNEI
Inter-Regional Academy of Personnel Management, Kyiv, Ukraine

HANS DADUNA
Hamburg University, Germany

PAVEL S. KNOPOV
V.M. Glushkov Institute of Cybernetics, National Academy of Science of
Ukraine, Kyiv, Ukraine

 Springer

Library of Congress Control Number: 2005938212

ISBN-10: 0-387-30409-6   e-ISBN: 0-387-31279-X

ISBN-13: 978-0387-30409-0

Printed on acid-free paper.

AMS Subject Classifications: 60G60, 60K15, 60K35, 90B15, 90C40, 91A15

Printed in the United States of America.

9 8 7 6 5 4 3 2 1

springer.com

CONTENTS

PREFACE

This book is devoted to the study and optimization of spatiotemporal stochastic processes. These are processes that simultaneously develop in space and time under random influences. Such processes occur almost everywhere when the global behavior of complex systems is studied, e.g., in physical and technical systems, population dynamics, neural networks, computer and telecommunication networks, complex production networks and flexible manufacturing systems, logistic networks and transportation systems, environmental engineering, climate modeling and prediction, earth surface models, and so on.

In the study of spatiotemporal stochastic processes the classical concepts of random fields (which are models for spatially distributed random phenomena) and of stochastic processes (hich are usually thought to describe the evolution over time of systems under random influences) converge. Over the last twenty years,many research monographs were written with emphasis on this unifying point of view, as were a huge number of articles and papers on this subject.

In many models, the space carries an additional structure. Some important cases are determined by geographical structures on some sphere, or by underlying regular lattices that determine an ordering of coordinate spaces in models of statistical physics, or by general graphs that determine network state spaces, say of the Internet. Random fields as models for spatially distributed systems provide us with a well established theory for all these specific variants of state spaces for a system under observation. The distribution of a random field contains the information on the random fluctuations of the spatial systems for a fixed time point. A realization of such a random field is therefore a snapshot of some system at a fixed instant in time. The mathematical theory and the statistics for random fields is an important branch of ongoing research.

For all the state spaces considered in this book, we assume that there is an underlying graph structure that reflects the interaction of the coordinates, which are associated to the vertices of the graph. This interaction graph determines a local structure for the state space of the systems, the edges of the graph determine neighborhoods within the space. Such spatial neighborhood systems also have a long history in systems that carry a regular lattice structure as, for example, in the standard models of statistical physics (Ising model, spin glasses) or in systems over non-regular graphs in stochastic networks (ARPANET, queueing networks, migration networks).

Important subclasses of random fields with structured state space are Gibbs fields (which were developed in physics as equilibrium states for models from statistical mechanics) and Markov random fields (which where developed in analogy to the structure of Markov processes in time), which means that the future depends on the past only through the present state. Astonishingly enough, it turned out that both notions, which elaborate on the nearest neighborhood concept, are essentially equivalent. The essence of both concepts can be summarized in view of our program: The statistical properties of any node depend on the global status of the system only through the local states within the neighborhood.

The importance of this concept cannot be overestimated, especially with the emergence of really large networks that evolve in time. There it is usually completely impossible to gather information of the global system state for possible decision making at a local level. So, it is necessary to define and investigate mathematical models that enable us to reduce the complexity of decision making on the basis of global information to decision making in local structures. Clearly, this is not always possible when globally optimal decision rules are required and the system dynamics depend on the complete history of the system in space and time.

But it is the hope that such reduction of complexity is possible if the dynamics of the uncontrolled system exhibits a local structure from its very definition. Traditionally, a local structure of a process in time is the Markov property and a local structure in space is the Gibbs property, i.e., the Markov random field property.

Our central topic is therefore to investigate systems that are Markovian in time and where the dynamics, determined by the transition operators, show a structure that resembles the Markov field property in space. The aim is to show that in many cases, optimization of these systems can be done by considering only strategies that respect the neighbourhood structure and the Markov property in time. This means that under specified conditions, we are able to exploit the locality of the systems' internal structure to find a set of rules we can use to establish a control regime that needs for decision making at a specific node (process coordinate) only information from the nearest neighbour nodes. We believe that this will help to reduce complexity of control schemes — although we sometimes pay by obtaining globally suboptimal policies that are only optimal in a set of local policies.

Our framework is classical stochastic dynamic optimization and it is well known that in many cases, especially if the underlying dynamics of the uncontrolled system is Markov in time, it is provable that optimal strategies can be found in subclasses of the class of admissible policies, where the history of the system is irrelevant for decision making if the present state is known. Such policies are called Markovian, adequately in line with the denominations above.

To summarize, the project undertaken in this book is to establish optimality or nearly optimality for Markovian policies in the control of spatiotemporal Markovian processes. We apply this general principle to different frameworks of Markovian systems and processes. Depending on the structure of the systems and the surroundings of the model classes, we arrive at different levels of simplicity for the policy classes that encompass optimal or nearly optimal policies. We accompany these theoretical findings by a set of examples that will hopefully demonstrate that there are important application areas for these theorems.

ACKNOWLEDGMENTS

We are grateful to the Scientific Editor of this book, Professor Panos Pardalos, and the Senior Publishing Editor, John Martindale, for their helpful support and collaboration in preparing the manuscript.

We thank our colleagues from the V. M. Glushkov Institute of Cybernetics of the National Academy of Sciences of Ukraine and the Center of Mathematical Statistics and Stochastic Processes at the Department of Mathematics of Hamburg University for many helpful discussions on the problems and results described and presented in this book.

We thank two anonymous referees for their careful reading of the text and for many helpful remarks.

We thank the German Research Association (Deutsche Forschungsgemeinschaft), who funded our collaboration by enabling the third author to stay with the Center of Mathematical Statistics and Stochastic Processes at the Department of Mathematics of Hamburg University for several long-term visits.

Chapter 1

INTRODUCTION

With the emergence of highly structured complex network structures in various fields of sciences (for example in physical and technical systems, population dynamics, neural networks, computer and telecommunication networks, climate modeling and prediction, earth surface models, environmental engineering, complex production networks and flexible manufacturing systems, logistic networks, and transportation systems) the research on spatiotemporal stochastic processes as models for such systems has found much interest in mathematics, econometrics, and computer science.

Spatiotemporal stochastic processes

Spatiotemporal stochastic processes are processes that simultaneously develop in space and time under random influences. In a first attempt, the term *spatiotemporal* stochastic process does not add really new information and meaning to the definition of a classical stochastic process that is a mathematical model for any system developing in time and the actual state of which is an element of its associated state space. As a typical example, the multidimensional Poisson processes or Brownian motion fit such a description.

Nevertheless, during the last twenty years, many research monographs were developed that put emphasis on the parallel evolution of systems in space and time, as were a huge number of articles and papers on the subject. Usually in such studies the state space carries an additional structure. Some important cases are determined by geographical structures on some sphere, or by an underlying regular lattice that determines an ordering of coordinate spaces in models of statistical physics, or by nonregular graphs that determine network state spaces, say of the Internet. We may comment on this by saying that in the study of spatiotemporal stochastic processes, the classical concepts of random fields (which are models for spatially distributed random phenomena) and of stochastic processes (which are usually thought to describe the evolution over time of systems under random influences) converge.

Before proceeding with a general discussion, we describe some examples that we consider to be prototypes of spatiotemporal stochastic processes.

As a first example, we consider a network of locations (say, villages or towns for human beings, or colonies of animals of the same or some interacting species) that interact by migration of the inhabitants. Species usually attempt to migrate for a variety of individual or population reasons and there is an additional development of the population by birth-death occurrences. For a more in-depth discussion, see the introduction of Chapter 9, *Spatial Population Dynamics,* in Renshaw's book MODEL-ING BIOLOGICAL POPULATIONS IN SPACE AND TIME [Ren93], and the examples that follow.

A second example is provided by Christakos' book RANDOM FIELD MODELS IN EARTH SCIENCES [Chr92], where Chapter 5 is devoted to *The Spatiotemporal Random Field Model.* The author strongly advocates to incorporate into the classical random field models that describe the state of an earth science scenario at a fixed time point also the development of these scenarios over time (for example, to assess the spatiotemporal variability of the earth's surface temperature and to predict extreme conditions).

The difference between the state spaces in these classes of models is obviously that the population dynamics are connected with local states on a grid or lattice (which represents the locations) or a graph (which represents the locations and the paths between the locations) while the temperature is usually defined over a continuous area. (Unless we consider the measure points as a predefined grid for the describing process.)

The importance of the general random field notion is that it provides us with models for spatially distributed systems. There is now a well established structure theory as well as a statistical theory for random fields, which is often summarized as in the title SPATIAL STATISTICS of Ripley's short book [Rip81], or in Cressie's STATISTICS FOR SPATIAL DATA [Cre99].

When considering dynamics associated with such distributed systems, the probability law of a random field contains the information on the random fluctuations of the spatial system for a fixed time point. A realization of such a random field is therefore a snapshot of some system at a fixed instant of time. The mathematical theory and the statistics for random fields is an important branch of ongoing research.

The state space of a random field is usually a product space $X = E^T$, where T is an index set that represents the sites or locations of the random field, while E can be considered as the attributes associated with the sites. A main distinction for random fields is according to whether T is discrete or continuous; see the remark following our first examples, above. Both subclasses have many important applications and a rich mathematical theory, and in both subclasses, there is much interest on so-called Markov random fields and Gibbs fields.

Local structures in spatiotemporal processes

In our investigations we always assume that there is some grid or graph that determines a discrete internal structure of the system's state space. A short introduction into migration processes that live on state spaces $\mathbb{N}^J$, where J is the number of colonies, is included in Chapters 2 and 6 of [Kel79]. The title of Kelly's book: REVERSIBILITY AND STOCHASTIC NETWORKS, introduces another connection to spatiotemporal processes. Stochastic networks or queueing networks have developed over almost thirty years from early applications in production systems and computer and telecommunications networks to a well established theory, which is still developing. For recent surveys, see [CY01] or [Ser99]. In queueing networks, the routing of the items (customers) inside the network is usually described by some Markovian routing mechanism that immediately leads to a neighborhood graph by considering the transition graph of the routing chain. Migration processes can be considered as special queueing networks; see [Kel79] and Whittle's book SYSTEMS IN STOCHASTIC EQUILIBRIUM [Whi86]. In their books, Whittle and Kelly discuss related processes, called *Spatial Processes* [Kel79, Chapter 9]. We discuss these processes and related stochastic network processes in some detail in Section 5.2.1. An important property of the Markovian spatial processes is that their transition operator has a local structure, determined by a neighborhood graph and that under some additional conditions, the equilibrium distribution of the process is a Markovian random field with respect to this neighborhood graph.

Markov random fields and related structures are another area that is of central interest in our investigations. These are special random fields where the distribution of the random field is connected with neighborhood structures and boundary structures of the underlying state space. In a first attempt, they can be considered as a spatial counterpart of the Markov property in time of standard stochastic processes.

For the case of continuous index set $T = \mathbb{R}^n$, say, the Markov property is defined with respect to a prescribed set of surfaces $\partial D \in \mathfrak{D}$, each of which partitions $\mathbb{R}^n$ into three disjoint sets $D^+ + D^- + \partial D = \mathbb{R}^n$ such that the attributes at sites in D^+ and D^- are conditional independent given the attributes in (a small neighborhood of) ∂D. For more details in this direction, see [MY66, MY67, Yad80, Geo88, Roz81] (with English translation [Roz82]).

Our focus in this book is on the class of random fields with discrete index set T, which is usually the set of vertices V of a graph $\Gamma = (V, B)$. The edges in B then define the neighborhoods of the Markovian random field. If we denote by $N(k) = \{j \colon \{k, j\} \in B\}$ the neighborhood of node $k \in V$, then the Markov property is defined with respect to these neighborhoods such that the attributes at site k and at sites in $V - (\{k\} \cup$

$N(k))$ are conditionally independent given the attributes at $N(k)$.

For details in the random field context, see [Dob68, Dob71, Lig85, Lig99, Pre74, Geo88] and the collections [DKT78, DS78] (English translation [DS80]).

Although we emphasize the Markov property in space, it should be clear that there is a close connection to classical random fields from statistical physics where the formalism of Gibbs fields introduces locality into the model description. Gibbs fields are probability laws on a (regular) graph structured state space that describe the fluctuations of (large) particle systems in equilibrium [Dob68, Dob71, Geo88]. From the assumption to be in equilibrium in the classical theory, the time development seems to play a minor role.

These models of systems in equilibrium have been used in various other areas in the meantime, e.g., modeling information diffusion in spatially distributed populations, image representation and segmentation [Win95], and fluctuation of markets in equilibrium.

But the restriction on time invariant probability laws as an assumption on the models becomes more and more questionable in these areas of applications and there are now many investigations published on the time development and equilibrium behaviour of so-called interacting processes [DKT78, Lig85, Lig99, Pre74]. Most of these investigations concentrate on Markovian interacting processes, where the term Markov refers firstly to the Markov property in time. It is an immediate question under which conditions a Markov interacting process possesses as (one dimensional in time marginal) limiting or equilibrium distribution a Gibbs or Markov random field. Indeed, the first approach was more in line with the question, which processes show the classical Gibbs ensembles as stationary distribution; see [Lig85], where stochastic nearest neighborhood particle systems (Ising models and general spin systems) and their Markov processes are investigated, as well as contact processes and voter processes with nearest neighborhood structure.

The recent book of Voit's puts (among others) these models under the title THE STATISTICAL MECHANICS OF FINANCIAL MARKETS [Voi03]. Most parts of the book are on random walks and scaling limits, so merely on the macroscopic behavior of the market models. But there is also a chapter on *Microscopic Market Models* that fit in the class of models we are working on. The author considers (discrete) populations of agents who buy and sell units of one stock and by observing the prices over time, they adjust their trading decisions and behavior. Depending on the assumptions, these agents behave Markovian in time, or act with a longer memory. With Markovian behavior for some examples, a behaviour occurs that resembles Ising models or spin glass models. We believe that this is a promising class of models in this area of applications, especially when the restriction on being close to the original models of statistical physics is relaxed. Then we are able to obtain more versatile models with more general state spaces in the flavor of Kelly's spatial processes or in the realm of general stochastic networks. If this is successful, then the question of optimization obviously enters the stage.

Optimization and local structures

Clearly, optimization of interacting systems is an important topic whenever complex systems are driven by some control regime. So, our project is a continuation of many previous investigations. But what we have in the center of our investigations seems to be new: The strong emphasis on locality in the models and in the optimization procedures.

We believe that the importance of locality and the subsequent structural properties of the systems cannot be overestimated, especially with the emergence of really large networks that evolve in time. There it is usually completely impossible to gather information of the global system state for decision making at a local level. So it is necessary to define and investigate mathematical models that enable us to reduce the complexity of decision making on the basis of global information to decision making in local structures. Clearly, this is not always possible when globally optimal decision rules are required and the system dynamics depend on the complete history of the system in space and time.

We will show that such reduction of complexity is possible if the dynamics of the uncontrolled system exhibit a local structure from its very definition. Traditionally, a local structure of a process in time is the Markov property and a local structure in space is the Gibbs property and their relatives, i.e., the Markov random field property.

We undertake the planned investigations in several steps, which are concerned with different classes of systems — in parallel to the distinction usually made by classifying stochastic processes: Discrete and continuous time, discrete and continuous state space. Our starting point in every chapter is an uncontrolled Markov process (in Chapter 5.1, a semi-Markov process), that describes the interaction of sites or nodes with possibly node specific states in a way that the transition operator resembles Markovian random field structures. In the definition, we follow the construction of Vasilyev and others [Vas78]:

Given some structure determining graph where the vertices are the interacting nodes of our system and the edges represent the nearest neighborhood connections, then the present state of the system determines the transition to the next state at any site only through the states of the sites in the nearest neighborhood (*local transition probabilities*). Given this information in discrete time systems, the individual transitions at the sites are conditionally independent and occur at the same time (*synchronous transition probabilities*). For continuous time processes, we assume that the embedded jump chains have *local* and *synchronous transition probabilities*.

At this point, we are faced with a well known problem: Even if an ergodic Markov process possesses local and synchronous transition kernels with respect to a prescribed neighborhood structure, the limiting and stationary distribution is not necessarily a Markov field with respect to this neighborhood system. This means that although the one-step transitions are strictly local, the influences generated by some node may nevertheless diffuse to far away sites. It has been a field of intense research whether the steady state of a process with local and synchronous transition kernels is a Markov random (Gibbs) field with respect to the same neighborhood structure. We discuss some positive results for discrete time processes from the literature in Section 3.2 and for continuous time processes in Section 5.2.1. Fortunately enough, it turns out that we can prove our main results without the assumption that the limiting distributions of the controlled systems are Markov fields in the classical sense. Nevertheless, such a closure property would be welcomed in system evaluation because it enhances performance evaluation and makes the structural behavior of the system more smooth.

For Markovian (or semi-Markovian) processes with local and synchronous transition kernels, we define *local* and *synchronous strategies* (*policies, plans*) to control the behavior of the system such that the local space structure of the system is respected under control. We refer to these strategies often simply as *local strategies*. These local strategies are, in general, randomized strategies and depend on the whole history of the system. This enables a decision maker to include his former experience with the system's reactions into decision making, i.e., to learn about the system's behavior.

In the framework of classical stochastic dynamic optimization, it is well known that in many cases, especially if the underlying dynamics of the uncontrolled system is Markov in time, it is provable that optimal strategies can be found in subclasses of the class of admissible policies, where the history of the system is irrelevant for decision making if the present state is known. Such policies are called Markovian as well. Often it can even be proved that there are optimal Markovian policies that furthermore, are deterministic decisions.

The Markov property here is with respect to time development, but in the classical optimization procedures with respect to space, we usually need the complete information about the actual state for decision making via a Markovian strategy. In general, there seems to be no hope to find optimal local strategies in the class of all admissible (including global) strategies.

But possibly this is not a really serious drawback in many situations. In large networks where decision makers have access to only a limited number of neighbored nodes, policies that depend on global information are not executable, so they cannot be admissible. The problem we are faced with is then:

Suppose we consider only local policies — are there in this class optimal policies that are Markovian in time as well?

Discussion of this problem in the different settings and proving the existence of optimal Markovian (in time) policies in the class of local policies is our main business in all the chapters that follow. Our main result is throughout Chapters 3 and 5 to provide conditions that guarantee that in specified local classes of policies Markovian policies exist, maybe under some additional conditions, e.g., on smoothness of the transition operators.

It should be obvious that such a procedure will not lead to proving optimality of Markovian local strategies in the class of all admissible (global) policies, but only to suboptimal strategies. But under the observation that decision making on the basis of global information is impossible, our *local search* will usually be the best that we can do.

Nevertheless, in special situations local optimization may be globally optimal. We discuss this in Sections 3.4.5 and 3.5.2. In both cases, we assume that the reward functions are separable, i.e., they are sums of locally defined reward functions, and these latter depend on the state of the nearest neighbors only.

Stochastic games and their mathematical investigation are closely related to stochastic dynamic optimization. We apply our principles in Chapter 4 to stochastic sequential multiperson games with an infinite horizon. The players are distributed on a graph and act locally.

Chapter 6 is devoted to some problems related to the spatiotemporal processes that we considered in earlier chapters: These problems are from different areas, e.g., information diffusion, image recognition and classification using Markov random field models, parameter estimation in random fields, and financial markets. These examples will illustrate the principles that guided us through the previous chapters, although there are in any case differences to our modeling principles.

We want to emphasize that the results obtained in our research, which are gathered here, cannot be a complete study of all problems connected to optimization of spatiotemporal processes that are Markovian in space and time. We therefore tried to list in the Bibliography articles and books that we consider to be related to the problems investigated in this book.

Notation:

We define sums with an empty index set to 0, and products with an empty index set to 1.

$\circ$ denotes the composition of functions: Let $f\colon A \to B$ and $g\colon B \to C$ then $f \circ g\colon A \to C$ is defined by $f \circ g(x) = f(g(x))$, $x \in A$.

Throughout the book we always assume that a suitable underlying probability space $(\Omega, \mathcal{F}, \mathrm{Pr})$ is given where all random variables are defined. At some points we will modify $(\Omega, \mathcal{F}, \mathrm{Pr})$, which will be indicated by a dedicated notation.

For any set X, we denote by 2^X the set of all subsets of X.

For any nonempty set X and any subset $\mathcal{E} \subseteq 2^X$, we denote by $\sigma_X(\mathcal{E}) = \sigma(\mathcal{E})$ the sigma algebra over X generated by $\mathcal{E}$.

If $(X, \mathfrak{X})$ is a measurable space and $A \in \mathfrak{X}$, then the indicator function of A is $\mathbb{1}\colon (X, \mathfrak{X}) \to \{0, 1\}$ with

$$\mathbb{1}_A(x) = \begin{cases} 1, & \text{if } x \in A, \\ 0, & \text{if } x \notin A. \end{cases}$$

We occasionally use the indicator function as logical value by writing

$$\mathbb{1}(A) = \begin{cases} 1, & \text{if the condition (event) } A \text{ is in force,} \\ 0, & \text{otherwise.} \end{cases}$$

Chapter 2

PREREQUISITES FROM THE THEORY OF STOCHASTIC PROCESSES AND STOCHASTIC DYNAMIC OPTIMIZATION

In this chapter we collect some fundamental notions for stochastic processes needed throughout the text and formalize the notions of a decision model in discrete and in continuous time. For the latter we follow closely the presentation in [Hin70] and [GS79].

2.1 STOCHASTIC PROCESSES

Definition 2.1 (Stochastic process). A family of random variables $\xi = \left(\xi^t\colon t \in T\right)$, in more detail:

$$\xi = \left(\xi^t\colon (\Omega, \mathcal{F}, \Pr) \to (X, \mathfrak{X}), t \in T\right)$$

with $T \neq \emptyset$ is called a stochastic process. Here $(\Omega, \mathcal{F}, \Pr)$ is the underlying probability space and $(X, \mathfrak{X})$ is the state space of the process.

 If the state space X is discrete (countable), then we always assume $\mathfrak{X} = 2^X$. $\odot$

The index set T has many relevant interpretations. If $T \subseteq \mathbb{R}$, then T is often interpreted as time, which may be discrete, $T = \mathbb{N}, \mathbb{Z}$ or subsets thereof, or continuous $T = \mathbb{R}, [0, \infty)$ or subsets thereof. The classical theory of stochastic processes is concerned with this setting, and the systems we are mostly interested in evolve on one of these time scales.

If T is not linearly ordered, e.g., if $T = \mathbb{Z}^2$ is a regular lattice, then T is often interpreted as space. Again, space may be continuous or discrete; in either case, ξ is then often called a random field.

If $T = V$ is the set of vertices of some graph $\Gamma = (V, B)$ with edge set B, then the state space is structured by B and coordinates of the space, which are neighbors according to the edges of Γ, are thought to interact; for details we refer to the fundamental Definition 3.2, which lays the groundwork for interacting systems we are interested in.

Such a random field describes the global state of a system at some fixed time point, and ξ_j, for $j \in V$ records the actual local state of the system at vertex (location) $j \in V = T$.

If such system with an interaction structure that is thought to be varying in space is given, we may then equip this with an additional time scale, say $\mathbb{N}$, resulting in a stochastic process

$$\xi = \left(\xi_j^t \colon (\Omega, \mathcal{F}, \mathrm{Pr}) \to (X, \mathfrak{X}), (t, j) \in T = \mathbb{N} \times V \right). \qquad (2.1)$$

Such a process describes the evolution in space and time of a distributed system with interacting components.

Remark 2.2. We usually separate in the general notion (2.1) the time variables from the space variables by writing

$$\xi = \left(\xi^t \colon (\Omega, \mathcal{F}, \mathrm{Pr}) \to (X, \mathfrak{X}) = \left(\underset{i \in V}{\times} X_i, \underset{i \in V}{\bigotimes} \mathfrak{X}_i \right), t \in \mathbb{N} \right)$$

such that ξ^t is a V-indexed random vector and we allow the local states at the different nodes to be different. $\odot$

The most prominent class of stochastic processes we are interested in during the modeling process are Markov processes.

Definition 2.3 (Markov processes). Let

$$\eta = \big(\eta^t \colon (\Omega, \mathcal{F}, \mathrm{Pr}) \to (X, \mathfrak{X}), t \in T\big)$$

be a stochastic process with parameter set $T \subseteq \mathbb{R}$ and denote by

$$\mathcal{F}_t^0 = \sigma\{\eta^s \colon s \le t\}$$

the pre-t σ-algebra of η (the σ-algebra of the past of η before t), and by

$$\mathcal{F}_t' = \sigma\{\eta^s \colon s \ge t\}$$

the post-t σ-algebra of η (the σ-algebra of the future of η after t).
η is a Markov process if the following holds:
For all $t \in T$ and all $B \in \mathcal{F}_t'$, we have Pr-almost surely

$$\mathsf{E}\left[\mathbf{1}_B \mid \mathcal{F}_t^0\right] = \mathsf{E}\left[\mathbf{1}_B \mid \eta^t\right].$$

We always assume that there exists a family of regular transition probabilities for η, i.e., a family $\mathsf{P} = \big(\mathsf{P}(s,t) \colon s, t \in T, s \le t\big)$ of kernels

$$\mathsf{P}(s,t) \colon X \times \mathfrak{X} \to [0,1], \quad (x, C) \to \mathsf{P}(s,t;x,C)$$

such that for all $s \le t$ and all $x \in X$ and $C \in \mathfrak{X}$ holds

$$\mathsf{P}(s,t;x,C) = \mathrm{Pr}\left(\eta^t \in C \mid \eta^s = x\right).$$

η is a (time) homogeneous Markov process if the kernels $\big(\mathsf{P}(s,t) \colon s, t \in T, s \le t\big)$ depend on s and t only via $t - s$. We then write $\big(\mathsf{P}(s,t) = \mathsf{P}(t-s) \colon s, t \in T, s \le t\big)$ and have a family $\mathsf{P} = \big(\mathsf{P}(t) \colon t \in T\big)$ of kernels such that for all $t \in T$ with $s, t \in T$ and all $C \in \mathfrak{X}$

$$\mathsf{P}(t;x,C) = \mathrm{Pr}\left(\eta^{s+t} \in C \mid \eta^s = x\right)$$

holds. If $0 \in T$, we then have the usual relation

$$P(t; x, C) = \Pr\left(\eta^{s+t} \in C \mid \eta^s = x\right)$$
$$= \Pr\left(\eta^t \in C \mid \eta^0 = x\right).$$

We always assume that $P(0)$ is the identity operator.

We use the term *Markov process* to refer to homogeneous Markov processes. Exceptions will be explicitly noted. $\odot$

Throughout the text we assume that the state spaces are smooth enough to guarantee that in connection with conditional expectations, regular conditional probabilities exist and we shall use this without mentioning it further.

For more details on Markov jump processes, see Subsection 5.2.1.

Definition 2.4 (Markov chains). A (homogeneous) Markov process on discrete time scale (usually $T \subseteq \mathbb{Z}$) is called a (homogeneous) Markov chain. The probabilistic transition behavior of a Markov chain is determined by the onestep transition kernels $P(1; x, C)$. We therefore introduce for homogeneous Markov chains with time scale $\mathbb{N}$

$$\xi = \left(\xi^t \colon (\Omega, \mathcal{F}, \Pr) \to (X, \mathfrak{X}), t \in \mathbb{N}\right)$$

throughout the notation

$$\Pr\left(\xi^{t+1} \in C \mid \xi^t = x\right) = P(1; x, C)$$
$$= Q(x; C)$$
$$= Q(C \mid x), \quad t \in \mathbb{N},$$

for all $x \in X$ and $C \in \mathfrak{X}$ and similarly for other discrete time scales.

If the state space X is discrete (countable), then the onestep transition kernels are determined by the stochastic matrices of the respective transition counting densities. We use the same symbols for the kernels and the associated stochastic matrices. $\odot$

Definition 2.5. If for a Markov chain ξ with discrete state space and transition matrix $Q(y \mid x)$ some probability π exists which fulfills

$$\sum_{x \in X} Q(y \mid x)\pi(x) = \pi(y), \quad y \in X, \tag{2.2}$$

then π is called an invariant or steady state distribution of ξ. $\odot$

Definition 2.6 (Markov processes with discrete state space). If

$$\eta = \left(\eta^t \colon (\Omega, \mathcal{F}, \mathrm{Pr}) \to (X, 2^X), t \in (0, \infty) \right)$$

is a continuous time Markov process with discrete state space such that

$$\lim_{t \downarrow 0} \mathsf{P}(t) = \mathsf{P}(0)$$

holds, i.e., the family of transition kernels is standard, then the right derivative

$$\lim_{t \downarrow 0} \frac{1}{t} \left(\mathsf{P}(t) - \mathsf{P}(0) \right) = \mathfrak{Q}$$

is called the Q-matrix of $\mathsf{P} = \left(\mathsf{P}(t) \colon t \geq 0 \right)$. $\odot$

To exclude pathological behavior we enforce the following assumption.

Assumption 2.7. If the state space of a Markov process η is a topological space, then the paths of η are assumed to be right continuous with left-hand limits (cadlag paths).

For any homogeneous Markov process η with discrete state space $(X, 2^X)$, we assume throughout that its paths are right continuous with left-hand limits (cadlag paths), that its Q-matrix $\mathfrak{Q}$ is conservative, which means

$$\sum_{y \in X - \{x\}} q(x, y) = -q(x, x), \quad x \in X,$$

and that η is non explosive (having only a finite number of jumps in any finite time interval with probability one).

Unless otherwise specified, we assume that η is irreducible on X. $\odot$

Corollary 2.8. Let

$$\eta = \left(\eta^t \colon (\Omega, \mathcal{F}, \mathrm{Pr}) \to (X, 2^X), t \in (0, \infty) \right)$$

be a continuous time Markov process with discrete state space that fulfills the Assumption 2.7. Then η can be characterized uniquely by a sequence $(\xi, \tau) = \left\{ (\xi^n, \tau^n), n = 0, 1, \dots \right\}$, which describes the interjump times τ^n and the successive states ξ^n, which the process enters at the jump instants.

The sequence of jump times of η is $\sigma = \{ \sigma^n \colon n = 0, 1, \dots \}$, given by $\sigma^0 = 0$, and $\sigma^n = \sum_{i=1}^n \tau^i$, $n \in \mathbb{N}$, and therefore for $t \in [\sigma^n, \sigma^{n+1})$, we have $\eta^t = \xi^n$, $n \in \mathbb{N}$.

The sequence $\xi = \left\{ \xi^n = \eta^{\sigma^n}, n = 0, 1, \dots \right\}$ is a homogeneous Markov chain, called the embedded jump chain of η. The one-step transition probability of the embedded jump chain is the Markov kernel

$$Q(y \mid x) = \mathrm{Pr} \left(\xi^{n+1} = y \mid \xi^n = x \right)$$
$$= \frac{q(x, y)}{-q(x, x)}, \quad x, y \in X. \qquad \odot$$

The definition of an embedded jump chain carries over to the case of general Markov jump processes; see the detailed description in Definition 5.19.

One of the possibly most important examples of processes with discrete state space is a birth-death process, for details see Theorem 5.23. Birth-death processes will serve as building blocks of the network processes and migration processes that we describe in Section 5.2.1.

Definition 2.9 (Birth-death processes). Let

$$\xi = \left(\left(\xi^t \colon (\Omega, \mathcal{F}, \mathrm{Pr}) \longrightarrow (\mathbb{N}, \mathcal{P}(\mathbb{N})) \right) \colon t \in \mathbb{R}_+ \right)$$

denote a Markov process with right continuous paths having left-hand limits (cadlag paths) and Q-Matrix $\mathfrak{Q} = \big(q(m,n)\colon m, n \in \mathbb{N}\big)$ given by

$$q(m,n) = \begin{cases} \lambda(m), & \text{if } 0 \le m, n = m + 1; \\ \mu(m), & \text{if } 1 \le m, n = m - 1; \\ -\lambda(0), & \text{if } m = n = 0; \\ -\big(\lambda(m) + \mu(m)\big), & \text{if } m = n > 0; \\ 0, & \text{otherwise.} \end{cases}$$

Then η is a (one dimensional) birth-death process with birth rates $\lambda(\cdot)$ and death rates $\mu(\cdot)$.

Unless otherwise specified, we assume $\mu(m) > 0$, $\forall\, m \ge 1$. $\lambda(m)$ may be 0 for some $m \in \mathbb{N}$. $\odot$

A class of processes that is often amenable to explicit structural investigation and computation of steady state distribution is the class of reversible Markov processes in continuous as well as in discrete time. For example, many of the processes that describe particle systems from statistical physics are reversible; see [Lig85].

Definition 2.10. A Markov chain $\xi = \big(\xi^t \colon t \in \mathbb{Z}\big)$ is called reversible (in time) if, for all $t \ge 0$ and $A, B \in \mathfrak{X}$, it holds

$$\mathrm{Pr}\left\{ \xi^t \in A, \xi^{t+1} \in B \right\} = \mathrm{Pr}\left\{ \xi^t \in B, \xi^{t+1} \in A \right\}. \qquad \odot$$

Lemma 2.11. (a) A reversible Markov chain is stationary.

(b) A Markov chain ξ with discrete state space is reversible if it is stationary and a strict positive probability measure π on X exists such that for all $x, y \in X$, we have

$$\pi(x)Q(y \mid x) = \pi(y)Q(x \mid y). \tag{2.3}$$

π is then the stationary probability of ξ. $\odot$

Reversibility of a Markov chain means that the operator defined by the onestep transition kernel which generates this Markov chain is self-adjoint; see [Str05, Section 5.1.1]. If this operator is symmetric or self-adjoint, in many cases it is easy to solve (2.2) for the steady state distribution by solving (2.3) instead. There are many cases where Markov chains are reversible. But (2.3) is a rather strong condition on the transition probabilities $Q(x \mid y)$.

The following criterion for reversibility of a Markov chain is of importance because it does not rely on having the stationary probability $\pi(x)$ explicitly given.

Theorem 2.12. Let the Markov chain ξ be stationary.

(a) Then ξ is reversible if and only if for all $n \geq 1$ and any sequence of states $x_1, \ldots, x_n \in X$ and $x_{n+1} = x_1$, we have

$$\prod_{i=1}^{n} Q(x_i \mid x_{i+1}) = \prod_{i=1}^{n} Q(x_{i+1} \mid x_i).$$

(b) The stationary probability π is then obtained as follows: Fix some $x_0 \in X$ and let

$$X^0 = \{x_0\}, \quad \text{and for } n \geq 0$$

$$X^{n+1} = \left\{ x \in X \backslash \bigcup_{k=0}^{n} X^k : Q(y \mid x) > 0 \text{ for some } y \in \bigcup_{i=0}^{n} X^i \right\}.$$

Set $\pi_{x_0} = 1$ and for $x \in X^{n+1}$, $y \in \bigcup_{k=0}^{n} X^k$ and $Q(y \mid x) > 0$

$$\pi(x) = \frac{Q(x \mid y)}{Q(y \mid x)} \pi(y),$$

and normalize finally. ⊙

Definition 2.13. A Markov process $\eta = (\eta^t \colon t \in \mathbb{R})$ in continuous time with discrete state space is called reversible (in time) if for all $n > 1$ and for all time points $t_1 < t_2 < \cdots < t_n$ and for all states $x_1, x_2, \ldots, x_n \in X$ holds

$$\Pr\left\{\eta^{t_i} = x_i \colon i = 1, \ldots, n\right\} = \Pr\left\{\eta^{s-t_i} = x_i \colon i = 1, \ldots, n\right\}. \quad ⊙$$

Lemma 2.14. (a) A reversible Markov process as given in Definition 2.13 is stationary.

(b) A Markov process η with discrete state space is reversible if it is stationary and a strict positive probability measure π on X exists such that for all $x, y \in X$ we have

$$\pi(x) q(x, y) = \pi(y) q(y, x).$$

π is then the stationary probability of η. ⊙

Corollary 2.15. A stationary birth-death process according to Definition 2.9 is a reversible Markov process.

If the embedded jump chain of the birth-death process is stationary then it is reversible. ⊙

2.2 DISCRETE TIME DECISION MODELS

Optimization of systems under stochastic influences is a challenging problem and is known to be often a complex operation. Especially if the real systems under investigation are large, a careful modelling process is needed. Therefore a precise definition of decision models is necessary. We start with decision making in discrete time systems, i.e., the system is observed only at discrete subsequent time points and decision making is allowed at these time points only.

Definition 2.16. A general decision model in discrete time (see [Hin70]) consists of the following items:
- A nonempty state space X that is endowed with a σ-algebra $\mathfrak{X}$.
- A nonempty action space A that is endowed with a σ-algebra $\mathfrak{A}$.
- A sequence $\left(\bar{H}^t\colon t \in \mathbb{N}\right)$ of admissible histories, where $\bar{H}^0 = X$, $\bar{H}^{t+1} = \bar{H}^t \times A \times X$ for $t \geq 0$. Each $\bar{H}^t$ (containing $2t+1$ factor sets) is endowed with the respective product σ-algebra $\bar{\mathfrak{H}}^t$.
- A sequence $A = \left(A^t\colon t \in \mathbb{N}\right)$ of set valued functions, which determines the admissible actions. $A^t\colon H^t \subseteq \bar{H}^t \to 2^A - \{\emptyset\}$, where the domain H^t is recursively defined as $H^0 := X$, and $H^{t+1} := \left\{(h, a, x) \in \bar{H}^{t+1}\colon h \in H^t, a \in A^t(h), x \in X\right\}$. $A^t(h)$ is the set of admissible actions at time t under history h.

H^t is endowed with the trace-σ-algebra $\mathfrak{H}^t := H^t \cap \bar{\mathfrak{H}}^t$.

We denote $K^t := \left\{(h, a)\colon h \in H^t, a \in A^t(h)\right\}$, and shall always assume that these sets contain the graph of a measurable mapping. K^t is endowed with the trace of the product-σ-algebra $\mathfrak{K}^t := K^t \cap \mathfrak{H}^t \otimes \mathfrak{A}$.
- An initial probability measure q^0 on $(X, \mathfrak{X})$ and a sequence $\mathsf{Q} = \left(\mathsf{Q}^t\colon t \in \mathbb{N}\right)$ of transition kernels, where $\mathsf{Q}^t\colon K^t \times \mathfrak{X} \to [0, 1]$ is the transition law of the system from time t to $t+1$.
- A sequence $r = \left(r^t\colon t \in \mathbb{N}\right)$ of $\mathfrak{K}^t$-$\mathbb{B}$ measurable reward functions $r^t\colon K^t \to \mathbb{R}$, where $r^t(h, a)$ is the reward obtained in the time interval $(t, t+1]$ if the history $h \in H^t$ is observed until time t and the decision then is $a \in A^t(h)$.   ⊙

The control of the decision model is performed by application of specified strategies to select under an observed history a decision variable that then triggers a new transition of the system's state. We will define different types of strategies that enable us to cover a variety of abstract problem formulations and real applications. For simplicity of presentation we first introduce deterministic strategies.

Definition 2.17. A deterministic admissible strategy (policy, control sequence, plan) is a sequence $\Delta = (\Delta^t \colon t \in \mathbb{N})$ of measurable functions $\Delta^t \colon X \to A$ with the following property:

If we use the strategy Δ and if up to time t the sequence of states occurred is $(x^0, x^1, \ldots, x^t)$ and the history observed is

$$h_\Delta \left(x^0, x^1, \ldots, x^t\right)$$
$$:= \left(x^0, \Delta^0(x^0), x^1, \Delta^1(x^0, x^1), \ldots, x^t, \Delta^t(x^0, x^1, \ldots, x^t)\right),$$

then we have

$$\Delta^t\left(x^0, x^1, \ldots, x^t\right) \in A^t\left(h_\Delta(x^0, x^1, \ldots, x^t)\right).$$

The functions Δ^t are called *decision rules, decisions,* or *actions*. We denote the set of all deterministic admissible strategies (policies, control sequences, plans) in a decision model by Π_P. (Deterministic admissible strategies are often called *pure* strategies.) ⊙

Definition 2.18. A randomized admissible strategy (control sequence, policy, plan) is a sequence $\pi = (\pi^t \colon t \in \mathbb{N})$ of transition kernels

$$\pi^t \colon H^t \times \mathfrak{A} \to [0,1]$$

from $(H^t, \mathfrak{H}^t)$ to $(A, \mathfrak{A})$, $h \in H^t$, such that for all histories $t \in \mathbb{N}$

$$\pi^t\left(h; A^t(h)\right) = \pi^t\left(A^t(h) \mid h\right) = 1$$

holds. (We use the notations $\pi^t(h; B) = \pi^t(B \mid h)$ as equivalent.)

We denote the set of all randomized admissible strategies (policies, control sequences, plans) in a decision model by Π. The transition kernels π^t are called *decision rules, decisions,* or *actions*.

A decision rule is called *Markovian* if for all $t \in \mathbb{N}$ and all histories $h = \left(x^0, a^0, x^1, a^1, x^2, \ldots, a^{t-1}, x^t\right)$, $g = \left(y^0, b^0, y^1, b^1, y^2, \ldots, b^{t-1}, y^t\right) \in H^t$ with $x^t = y^t$ we have $\pi^t(h; \cdot) = \pi^t(g; \cdot)$.

In such situation we call the strategy Markovian as well and consider a Markov strategy as a sequence $\pi = \left(\pi^t \colon t \in \mathbb{N}\right)$ of transition kernels $\pi^t \colon X \times \mathfrak{A} \to [0, 1]$ from $(X, \mathfrak{X})$ to $(A, \mathfrak{A})$.

We denote the set of all Markov (admissible) strategies in a decision model by Π_M.

Note that whenever we deal with Markovian strategies, we can assume that $A^t\left(h^t\right)$ depends on h^t only through x^t. We denote this restricted dependence by $A^t\left(h^t\right) =: A^t\left(x^t\right)$. This will be done without further mention.

A Markovian strategy is *stationary* if the transition kernels are time independent, i.e., $\pi^t = \pi^s$, $s, t \in \mathbb{N}$.

We denote the set of all stationary (Markovian admissible) strategies in a decision model by Π_S.

The set of all deterministic (pure) Markovian (admissible) strategies in a decision model is denoted by Π_{PM}.

The set of all deterministic stationary Markovian (admissible) strategies in a decision model is denoted by Π_D. $\odot$

Remark 2.19. Whenever we are dealing with deterministic strategies, we assume that all the involved σ-algebras contain the one-point sets. Then deterministic plans can be considered as randomized plans as well.

We then have

$$\Pi_D \subseteq \Pi_S \subseteq \Pi_M \subseteq \Pi$$

and

$$\Pi_D \subseteq \Pi_{PM} \subseteq \Pi_P \subseteq \Pi. \qquad \odot$$

Remark 2.20. We will later consider controlled processes in continuous time and use controls that are families $\pi = \left(\pi^t \colon t \geq 0\right)$ of suitable transition kernels as randomized controls, and pure strategies that are in the Markovian case then functions $\Delta = \left(\Delta^t \colon X \to A, t \in [0, \infty)\right)$.

Without further remarks we will use the same notation as in Remark 2.19.

The same procedure will apply if we are concerned with controlled processes in continuous time where the control and decision making is only allowed at an embedded sequence of random or deterministic time points, as, e.g., in the case of semi-Markov processes (Section 5.1) or Markov jump processes (Section 5.2). ⊙

If a decision model according to Definition 2.16 is given and a (randomized) admissible strategy according to Definition 2.18 is fixed then from the transition kernels $(Q^t : t \in \mathbb{N})$ for the state transitions and $(\pi^t : t \in \mathbb{N})$ for the decisions a dynamics for the system is specified over any finite time horizon $\{0, 1, \ldots, t\}$. We denote by

$$(\xi, \alpha) = \left((\xi^t, \alpha^t) : t \in \mathbb{N} \right)$$

the sequence of successive states and decisions and assume that this sequence is given for an infinite horizon. A consistent construction of a probability space $(\Omega, \mathcal{F}, \mathrm{Pr})$ where this stochastic process lives on can be done in the standard way by construction of the canonical process.

Let $\Omega = (X \times A)^{\mathbb{N}}$, $\mathcal{F} = (\mathfrak{X} \otimes \mathfrak{A})^{\mathbb{N}}$, α^t and ξ^t are the respective projections, and Pr is constructed with the help of the theorem of Ionescu Tulcea. The procedure is as follows.

For a prescribed (randomized) strategy π according to Definition 2.18, an initial distribution q^0 on $(X, \mathfrak{X})$ and sequence of transition kernels Q^t, $t \in \mathbb{N}$ (see Definition 2.16) we have for all $t \in \mathbb{N}$

$$\mathrm{Pr} \left\{ \xi^0 \in C^0, \alpha^0 \in B^0, \ldots, \xi^t \in C^t, \alpha^t \in B^t \right\}$$

$$= \int_{C^0} \mathsf{q}^0 \left(dx^0 \right) \int_{B^0} \pi^0 \left(x^0; da^0 \right) \int_{C^1} Q^0 \left(x^0, a^0; dx^1 \right) \times \cdots$$

$$\cdots \times \int_{C^t} Q^{t-1} \left(x^0, a^0, \ldots, x^{t-1}, a^{t-1}; dx^t \right) \times$$

$$\times \int_{B^t} \pi^t \left(x^0, a^0, \ldots, a^{t-1}, x^t; da^t \right) \quad (2.4)$$

with $C^s \in \mathfrak{X}, B^s \in \mathfrak{A}, s \leq t$. This sequence of finite dimensional distributions uniquely determines Pr and the distribution of (ξ, α), which is the sequence of the respective projections on Ω. Therefore (2.4) determines Pr on the cylindrical sets of $\mathcal{F} = (\mathfrak{X} \otimes \mathfrak{A})^{\mathbb{N}}$ by

$$
\begin{aligned}
\Pr \left\{ \xi^0 \in C^0, \alpha^0 \in B^0, \ldots, \xi^t \in C^t, \alpha^t \in B^t \right\} \\
= \Pr \left\{ (x^i, a^i) \colon x^i \in C^i, a^i \in B^i, i = 0, 1, \ldots, t \right\}.
\end{aligned}
$$

That Pr exists and is uniquely determined on $\mathcal{F} = (\mathfrak{X} \otimes \mathfrak{A})^{\mathbb{N}}$ is the result of Ionescu Tulcea.

It should be noted that formally we have to extend the domain of the $\mathbf{Q}^t$ from $K^t \times \mathfrak{X}$ to $(X \times A)^t \times \mathfrak{X}$. The construction sketched here is the most general one and does not need the assumption of having Polish state and action spaces. Moreover, if the strategy is deterministic, it is possible to construct an underlying probability space that governs the evolution of the decision model on some space with $\Omega = X^{\mathbb{N}}$, $\mathcal{F} = \mathfrak{X}^{\mathbb{N}}$; see [Hin70, page 80] or [GS79, Section 1.1].

We are faced with the problem of howto compare the behavior of different decision models with fixed transition mechanisms $\mathbf{Q}^t$, $t \in \mathbb{N}$, but under different initial distributions $\mathbf{q}^0$ and different strategies π. For easier reading, we will distinguish the different underlying probability measures by suitably selected indices in a form, say $\Pr_{\mathbf{q}^0}^\pi$, which in case that $\mathbf{q}^0$ is concentrated in the point x^0 will be abbreviated by $\Pr_{x^0}^\pi$. Expectations under $\Pr_{\mathbf{q}^0}^\pi$, will be written as $\mathsf{E}_{\mathbf{q}^0}^\pi$.

In our discrete time models the evaluation of the decision model and of the sequence $(\xi, \alpha) = \left((\xi^t, \alpha^t) : t \in \mathbb{N} \right)$, i.e., the assessment of the strategy and of the time behavior of the decision model, will be according to the asymptotic expected time average reward/costs principles. We will consider mainly cost and reward functions that are stationary in time, i.e., r^t is independent of t and therefore a function

$$
r = r^t \colon (X \times A, \mathfrak{X} \otimes \mathfrak{A}) \longrightarrow (\mathbb{R}_+, \mathbb{B}_+), \qquad \forall\, t \in \mathbb{N}. \qquad (2.5)
$$

Therefore, if at time $t \in \mathbb{N}$ the system is in state $\xi^t = x^t$ and a decision for action $\alpha^t = a^t$ is made a (one-step) cost $r\left(x^t, a^t\right) \geq 0$ is incurred to the system. The average expected cost up to time T when ξ is started with $\xi^0 = x^0$ and strategy π is applied is

$$\mathsf{E}^\pi_{x^0} \frac{1}{T+1} \sum_{t=0}^{T} r\left(\xi^t, \alpha^t\right),$$

where $\mathsf{E}^\pi_{x^0}$ is expectation associated with the controlled process (ξ, α) under π if $\xi^0 = x^0$.

The first problem is to find a strategy π that minimizes the maximal asymptotic average expected costs.

Definition 2.21. For the controlled process (ξ, α) under policy π and starting with $\xi^0 = x^0$ the asymptotic maximal expected time average cost is

$$\rho\left(x^0, \pi\right) = \limsup_{T \to \infty} \mathsf{E}^\pi_{x^0} \frac{1}{T+1} \sum_{t=0}^{T} r\left(\xi^t, \alpha^t\right). \qquad (2.6)$$

A strategy $\pi^\star \in \Pi$ is optimal with respect to the (minimax) cost criterion (in the class of admissible randomized strategies) if

$$\rho(x^0, \pi^\star) = \inf_{\pi \in \Pi} \rho(x^0, \pi), \qquad \forall \, x^0 \in X. \qquad \odot$$

The dual problem is to find a strategy π that maximizes the asymptotic average expected reward.

Definition 2.22. For the controlled process (ξ, α) under policy π and starting with $\xi^0 = x^0$ the asymptotic mainimal expected time average reward is

$$\phi\left(x^0, \pi\right) = \liminf_{T \to \infty} \mathsf{E}^\pi_{x^0} \frac{1}{T+1} \sum_{t=0}^{T} r\left(\xi^t, \alpha^t\right). \qquad (2.7)$$

A strategy $\pi^\star \in \Pi$ is optimal with respect to the (maximin) reward criterion (in the class of admissible randomized strategies) if

$$\phi\left(x^0, \pi^\star\right) = \sup_{\pi \in \Pi} \phi\left(x^0, \pi\right), \qquad \forall \, x^0 \in X. \qquad \odot$$

Whenever it is clear from the context whether we consider the (mini-max) reward criterion or the (maximin) reward criterion, we will use the phrase *optimal policy*.

2.3   CONTINUOUS TIME DECISION MODELS

In this section we consider stochastic processes with time scale $[0, T]$ or $[0, T)$ for $T \leq \infty$.

When studying controlled stochastic processes in continuous time, we often assume that the state spaces and the action spaces are Polish topological spaces that are endowed with Borel σ-algebras. This will provide sufficient generality for all the applications we have in mind and encompass the most prominent classes of stochastic processes used in applications. Our description in this introduction follows closely the presentation in [GS79].

Let $(X, \mathfrak{X})$ and $(A, \mathfrak{A})$ be measurable spaces, with σ-algebras $\mathfrak{X}$ and $\mathfrak{A}$. $(X, \mathfrak{X})$ is the state space of the basic stochastic process, and $(A, \mathfrak{A})$ is the action space for the control.

We denote by $X^{[0,T]}$ the space of all functions defined on $[0, T]$ with values in X and by $\mathfrak{X}^{[0,T]}$ the σ-algebra generated by cylinder sets from $X^{[0,T]}$. Further we define $A^{[0,T]}$ and $\mathfrak{A}^{[0,T]}$ in the same way for the measurable space $(A, \mathfrak{A})$.

Similarly, for $0 \leq s \leq t \leq T$, we define $\mathfrak{X}^{[s,t]}$ to be the σ-algebra over $X^{[0,T]}$ generated by cylinder sets with bases in $[s, t]$, and $\mathfrak{X}^{[s,t)}$ the

σ-algebra over $X^{[0,T]}$ generated by cylinder sets with bases in $[s,t)$. Further expressions for σ-algebras with other time intervals are to be read analogously.

Sometimes we abbreviate $\mathfrak{X}^t = \mathfrak{X}^{[0,t]}$ and $\mathfrak{X}^{t-0} = \sigma\left(\bigcup_{s<t}\mathfrak{X}^s\right) = \mathfrak{X}^{[0,t)}$. Similarly, we determine $\mathfrak{A}^{[s,t]}$, $\mathfrak{A}^{[s,t)}$, $\mathfrak{A}^t$ and $\mathfrak{A}^{t-0}$.

2.3.1 Continuous time decision models with step control

In this subsection we consider processes with time scale $[0,T]$, where $T \leq \infty$.

We follow in the next part the procedure of Gihman and Skorohod and define in a formal analogy to the discrete time situation a controlled object and a control as families of probability measures that resemble the definition of the respective transition kernels in discrete time and which may serve as similar objects in the continuous time setting.

Definition 2.23 (see [GS79]). A controlled object is a family of probability measures $\mu(C \mid a)$, defined for all events $C \in \mathfrak{X}^{[0,T]}$ and histories $a(\cdot) \in A^{[0,T]}$, which satisfies the following measurability condition:

For all $t \in [0,T]$ and all events $C \in \mathfrak{X}^{[0,t]}$ up to time t, the function $\mu(C \mid a(\cdot))$ is a $\mathfrak{A}^{[0,t)}$-measurable function of the second component $a(\cdot)$.

A control is a family of probability measures $\nu(B \mid x(\cdot))$ defined for all decision history events $B \in \mathfrak{A}^{[0,T]}$ and state space paths $x(\cdot) \in X^{[0,T]}$, which satisfies the following measurability condition:

For all $t \in [0,T]$ and all decision history events $B \in \mathfrak{A}^{[0,t]}$ up to time t, the function $\nu(B \mid x(\cdot))$ is a $\mathfrak{X}^{[0,t]}$-measurable function of the second component $x(\cdot)$. $\qquad\odot$

Note that we use the term *control* for the measure $\nu(\cdot \mid \cdot)$ as well as for the elements $a = a(\cdot) \in A^{[0,T]}$. The meaning will always be clear from the context.

In general, the construction of stochastic processes with given control and a controlled object is difficult.

The construction of associated processes is simpler for the case of deterministic controls, i.e., controls that are determined by a family of $\mathfrak{X}^{[0,t]}$-measurable functionals $\left\{\eta^t(x(\cdot)) : t \in [0,T]\right\}$, such that for $B \in \mathfrak{A}^{[0,t]}$, we have

$$\nu\big(B \mid x(\cdot)\big) = 1\!\!1_B\Big(\eta^t\big(x(\cdot)\big)\Big).$$

In this case for the controlled process $\big(\xi(t), \alpha(t)\big)$ the equality

$$\alpha(t) = \eta^t\big(\xi(\cdot)\big), \qquad 0 \le t \le T,$$

holds with probability 1, and hence it is possible to determine the control, although the controlled object ξ cannot be determined in this way.

This is because to construct the basic process on $[0,t]$, we need to know the control α on $[0,t)$, which in turn is only determined if the basic process ξ is known on $[0,t)$. In discrete time we have an iterative scheme to determine the process and the control step-by-step but, in continuous time, this is obviously not the case, for more details, see [GS79, page 80].

These problems will not occur if the control is *delayed* with respect to the process, i.e., knowledge of the control up to time t allows us to determine the state process on some time interval $[0, t + s]$, $s > 0$.

Definition 2.24. Let F be some nonempty set. A function $f : [0, \infty] \to F$ is a step function if it is piecewise constant and the sequence of jump points $0 = t^0 < t^1 < \cdots < t^n < \cdots$ of the function is either finite or diverges, i.e., $\lim_{n \to \infty} t^n = \infty$.

A function $f : I \longrightarrow F$ on a finite interval I is a step function if it is piecewise constant and the number of jumps of the function is finite.

We denote by $\bar{F}^{[0,T]}$ the set of all step functions $f : [0,T] \to F$.

A control $a(\cdot) \in A^{[0,T]}$ is a *step control* if it is a step function on $[0,T]$, i.e., for some sequence $0 = t^0 < t^1 < \cdots < t^n < \cdots$ holds

$$a(t) = a\big(t^k\big), \quad \forall\, t \in \big[t^k, t^{k+1}\big). \qquad\qquad \odot$$

Note that according to our definition, a step function has only isolated jump points.

Definition 2.25. A control $\nu(\cdot \mid \cdot)$ is a *step control* if for all $x(\cdot) \in X^{[0,T]}$ the measure $\nu(\cdot \mid x(\cdot))$ is concentrated on the set of all step functions in $A^{[0,T]}$. ⊙

Let for $x(\cdot) \in X^{[0,T]}$ a step control $\nu(\cdot \mid x(\cdot))$ be given. We describe next how to derive the details for the subsequent further process construction from this information.

Sojourn time distributions and jump probabilities:
Denote by $\sigma^1, \sigma^2, \ldots$ the random times at which the random control $\alpha(\cdot)$ changes its value. From the prescribed measure $\nu(\cdot \mid x(\cdot))$, we obtain the measure

$$\nu\Big(\{a(\cdot) \colon a(0) \in B^0\} \mid x(\cdot)\Big) =: \nu^0\big(B^0 \mid x(0)\big),$$

which determines the distribution of the control at the initial time 0. The distribution of the sojourn time in the initial state for the control is determined as follows: For all t

$$\nu\Big(\{a(\cdot) \colon a(0) \in B^0, a(s) = a(0), s \le t\} \mid x(\cdot)\Big)$$

is a $\mathfrak{X}^{[0,t]}$-measurable function. As a function of B^0 this measure is absolute continuous with respect to the measure $\nu^0(B^0 \mid x(0))$. This yields the existence of the density function

$$\lambda^1\big(t \mid x(\cdot), a(0)\big),$$

such that

$$\nu\Big(\{a(\cdot) \colon a(0) \in B^0, a(s) = a(0), s \le t\} \mid x(\cdot)\Big)$$
$$= \int_{B^0} \lambda^1\big(t \mid x(\cdot), a^0\big) \nu^0\big(da^0 \mid x(0)\big)$$

holds. $\lambda^1(t \mid x(\cdot), a(0))$ determines the probability that $\sigma^1 > t$ holds, i.e., the sojourn time distribution in the initial state of the control.

Given the initial state a^0 of the control $\alpha(\cdot)$ and the time $\sigma^1 = s^1$ of the first jump of the control, we further define

$$\nu^1(B^1 \mid x(\cdot), a^0, s^1),$$

the conditional probability that $\alpha(s^1) \in B^1$. Similarly, for all k, we define the density functions

$$\lambda^k(t \mid x(\cdot), a^0, \ldots, a^{k-1}, s^1, \ldots, s^{k-1}) \qquad (s^1 < \cdots < s^{k-1} < t)$$
$$\text{and} \quad \nu^k(B^k \mid x(\cdot), a^0, \ldots, a^{k-1}, s^1, \ldots, s^k) \qquad (s^1 < \cdots < s^k),$$

which are the conditional distributions of σ^k, respectively of $\alpha(\sigma^k)$, under the condition that

$$\sigma^1 = s^1, \ldots, \sigma^{k-1} = s^{k-1}, (\text{respectively } \sigma^k = s^k),$$
$$\text{and} \quad \alpha(\sigma^0) = a^0, \ldots, \alpha(\sigma^{k-1}) = a^{k-1}.$$

The functions λ^k and ν^k are measurable in all arguments, and $\lambda^k(t \mid \cdot)$ is measurable in $x(\cdot)$ with respect to $\mathfrak{X}^t$, and $\nu^k(\cdot \mid \cdot)$ is measurable with respect to $\mathfrak{X}^{[0,s^k]}$.

Analogously, starting from the controlled object μ, we introduce the conditional measures $\mu^t(C \mid x(\cdot), a(\cdot))$ defined on a σ-algebra $\mathfrak{X}^{[t,T]}$, measurable in $x(\cdot)$ and $a(\cdot)$ with respect to $\mathfrak{X}^{[0,t]} \times \mathfrak{A}^{[0,T]}$, and such that for any $C' \in \mathfrak{X}^{[0,t]}$ we have for the given control $a(\cdot)$

$$\int_{C'} \mu^t(C \mid x(\cdot), a(\cdot)) \mu(dx \mid a(\cdot)) = \mu(C \cap C' \mid a(\cdot)).$$

So the measure $\mu^t(C \mid x(\cdot), a(\cdot))$ determines the conditional distribution on $[t, T]$ of the processes corresponding to the measure $\mu(\cdot \mid a(\cdot))$ (the controlled object), if its value on $[0, t]$ and the control over $[0, T]$ are known.

Construction of the process distribution:

We now show how to determine for a given control and controlled object the distributions of the controlled process $\big((\xi(t), \alpha(t)) : t \in [0, T]\big)$ (which will be defined iteratively) by utilizing the functions λ^k, ν^k, and $\mu^t(\cdot \mid (\cdot), a(\cdot))$. In doing this we construct a sequence of processes (respectively their distributions)

$$\Big\{\big((\xi_n(t), \alpha_n(t)) : t \in [0, T]\big) : n \in \mathbb{N}\Big\}$$

as follows.

n=0: Given $\xi(0)$, the conditional distribution for $\alpha_0(0)$ is $\nu^0(B^0 \mid \xi^0)$, and for all $t \in [0, T]$ $\alpha_0(t) = \alpha_0(0)$ holds. Define $\xi_0(t)$ such that $\xi_0(0) = \xi(0)$ and for all $C \subset \mathfrak{X}^{[0,T]}$ holds

$$\Pr\{\xi_0(\cdot) \in C \mid \alpha_0(\cdot)\} = \mu\big(C \mid \alpha_0(0)\big).$$

n=1: We define σ^1 and $\alpha_1(\sigma^1)$ such that

$$\Pr\{\sigma^1 > t \mid \xi_0(\cdot), \alpha_0(\cdot)\} = \lambda^1\big(t \mid \xi_0(\cdot), \alpha_0(\cdot)\big),$$
$$\Pr\{\alpha_1(\sigma^1) \in B^1 \mid \xi_0(\cdot), \alpha_0(\cdot)\} = \nu^1\big(B^1 \mid \xi_0(\cdot), \alpha_0(\cdot), \sigma^1\big).$$

Put

$$\alpha_1(t) = \begin{cases} \alpha_0(t), & \text{if } t < \sigma^1; \\ \alpha_1(\sigma^1), & \text{if } t \geq \sigma^1, \end{cases}$$

and construct the process $\xi_1(t)$ such that $\xi_1(t) = \xi_0(t)$ holds for $t < \sigma^1$, and for $\sigma^1 \leq t$:

$$\Pr\{\xi_1(\cdot) \in C \mid \xi_1(s), s < t, \alpha_1(\cdot)\} = \mu^t\big(C \mid \xi_1(\cdot), \alpha_1(\cdot)\big), \quad C \in \mathfrak{X}^{[t,T]}.$$

n=k: Continuing this way, we define $\big(\xi_k(t), \alpha_k(t)\big)$ such that $\alpha_k(t)$ has exactly k jumps in $[0, T]$, say at jump times $0 < \sigma^1 < \cdots < \sigma^k$, and $\alpha_k(t) = \alpha_{k-1}(t)$ for $t < \sigma^{k-1}$, and $\xi_k(t) = \xi_{k-1}(t)$ for $t < \sigma^k$.

n=k+1: If we have constructed $\xi_k(t)$ and $\alpha_k(t)$, then we first determine the time instant σ^{k+1} and the value of $\alpha_{k+1}(\sigma^{k+1})$ such that

$$\Pr\left\{\sigma^{k+1} > t \mid \xi_k(\cdot), \alpha_k(\cdot)\right\}$$
$$= \lambda^{k+1}\left(t \mid \xi_k(\cdot), \alpha_k(0), \ldots, \alpha_k(\sigma^k), \sigma^1, \ldots, \sigma^k\right),$$
$$\Pr\left\{\alpha_{k+1}(\sigma^{k+1}) \in B^{k+1} \mid \xi_k(\cdot), \alpha_k(\cdot)\right\}$$
$$= \nu^{k+1}\left(B^{k+1} \mid \xi_k(\cdot), \alpha_k(0), \ldots, \alpha_k(\sigma^k), \sigma^1, \ldots, \sigma^k\right).$$

Then we set $\alpha_{k+1}(t) = \alpha_k(t)$ for $t < \sigma^{k+1}$, and $\alpha_{k+1}(t) = \alpha_k(\sigma^{k+1})$ for $t \geq \sigma^{k+1}$. If the process $\alpha_{k+1}(t)$ is constructed, we determine the process $\xi_{k+1}(t)$ by setting it to be equal to $\xi_k(t)$ on $[0, \sigma^{k+1}]$, and extending it to $[\sigma^{k+1}, T]$ such that for all $C \in \mathfrak{X}^{[t,T]}$ and $\sigma^{k+1} \leq t$ we have

$$\Pr\left\{\xi_{k+1}(\cdot) \in C \mid \xi_{k+1}(s), s < t, \alpha_{k+1}(\cdot)\right\} = \mu^t(C \mid \xi_{k+1}(\cdot), \alpha_{k+1}(\cdot)).$$

If $\sigma^{k+1} \geq T$, then the process $\left((\xi_{k+1}(t), \alpha_{k+1}(t)) \colon t \in [0, T]\right)$ is (for the path under construction) the required process $\left((\xi(t), \alpha(t)) \colon t \in [0, T]\right)$. (Note that the running index k is random.)

We have shown how to construct the controlled process from the measures μ and ν, where ν is a step control, such that the controlled process $\left((\xi(t), \alpha(t)) \colon t \in [0, T]\right)$ has the required control and controlled object. We will need te following explicit definition of a general construction related to that procedure. The definition recalls the previous construction starting from a given abstract process. We discuss this below on page 34

Definition 2.26 (Representation of a controlled object). Let $(\Omega, \mathcal{F}, \Pr)$ be a probability space. We say that the family of random processes $\xi(t, \omega; a(\cdot))$, $t \in [0, T]$, $\omega \in \Omega$, $a(\cdot) \in A^{[0,T]}$, is the representation of a controlled object $\mu(\cdot \mid \cdot)$ if the following conditions hold:

1) for all $C \in \mathfrak{X}^{[0,T]}$

$$\Pr \left\{ \xi(\cdot, \omega; a(\cdot)) \in C \right\} = \mu(C \mid a(\cdot));$$

2) if $a^1(t) = a^2(t)$ for all $t \leq \sigma^1$, then we have $\xi(t, \omega; a^1(\cdot)) = \xi(t, \omega; a^2(\cdot))$ for all $t \leq \sigma^1$;

3) $\xi(\cdot, \omega; a(\cdot))$ is a measurable random function defined on $A^{[0,T]}$ with values in $X^{[0,T]}$, i.e., for all $t \in [0, T]$ and $C \in \mathfrak{X}^{[0,T]}$ we have

$$\left\{ (\omega; a(\cdot)) \colon \xi(\cdot, \omega; a(\cdot)) \in C \right\} \in \mathcal{F} \times \mathfrak{A}^{[0,t]}.$$

Denote by $(\mathcal{F}^t \colon t \in [0, T])$ the natural filtration in $\mathcal{F}$, generated by ξ, i.e., $\mathcal{F}^t = \sigma \left\{ \xi(s, \cdot, a(\cdot)) \colon s \leq t, a(\cdot) \in A^{[0,T]} \right\}$.

A generalized control is an arbitrary process $\alpha = (\alpha(t) \colon t \in [0, T])$ with values in A for all t, which is measurable with respect to $\mathcal{F}^{t-0} = \sigma \left(\bigcup_{s < t} \mathcal{F}^s \right)$. $\odot$

The first condition in Definition 2.26 is necessary for the process $\xi(\cdot, \omega; a(\cdot))$ to have for fixed $a(\cdot)$ the same distributions as the controlled process with the controlled object $\mu(\cdot \mid \cdot)$ and the fixed control $a(\cdot)$. The second condition is a consistence condition for the control and the basic process: To define the basic process on $[0, t]$, it is necessary to define the control on $[0, t]$. The third condition is necessary to replace $a(\cdot)$ in $\xi(\cdot, \omega; a(\cdot))$ by the process $\alpha(\cdot)$.

Remark 2.27. Under condition 3) from Definition 2.26, it follows that for any generalized control $\alpha(\cdot)$ the process $(\xi(t, \omega), \alpha(\cdot))$ is a random process on $(\Omega, \mathcal{F}, \Pr)$. We call it the controlled process under control $\alpha(\cdot)$.

If instead of condition 3) of Definition 2.26 the following condition 3′ holds:

3′) $\left\{ (\omega, a(\cdot)) \colon \xi(\cdot, \omega, a(\cdot)) \in C \right\} \in \mathcal{F}^t \times \mathfrak{A}^{[0,t]}$, for all $t \in [0, T]$, $C \in \mathfrak{X}^t$, then the process $\xi(t, \omega, \alpha(t))$ is even $\mathcal{F}^t$-measurable. $\odot$

In the following we study representations of controlled objects that are of the structure given in Definition 2.26. We restrict ourself to step controls, and additionally assume that the controlled object has paths that are step functions, i.e., for all $a(\cdot) \in A^{[0,T]}$ the probability measure $\mu(\cdot \mid a(\cdot))$ fulfills $\mu\left(\bar{X}^{[0,T]} \mid a(\cdot)\right) = 1$, where $\bar{X}^{[0,T]}$ is a set of all step functions in $X^{[0,T]}$. A step controlled object can be defined by the following set of conditional distributions:

$$\mathsf{P}^0\left(dx^0\right), \quad \lambda^1\left(ds \mid x^0; a(\cdot)\right), \quad \mathsf{P}^1\left(dx^1 \mid x^0, t^1; a(\cdot)\right), \quad \ldots,$$
$$\lambda^k\left(ds \mid x^0, \ldots, x^{k-1}, t^1, \ldots, t^{k-1}; a(\cdot)\right),$$
$$\mathsf{P}^k\left(dx^k \mid x^0, \ldots, x^{k-1}, t^1, \ldots, t^k; a(\cdot)\right), \quad \ldots, \tag{2.8}$$

where $\mathsf{P}^0\left(dx^0\right)$ is the distribution of $x(0)$, which is independent from $a(\cdot)$; $\lambda^1\left(ds \mid x^0; a(\cdot)\right)$ is the conditional distribution of the first jump time of the process; $\mathsf{P}^1\left(dx^1 \mid x^0, t^1; a(\cdot)\right)$ is the conditional distribution of the state of the process after the first jump, given the time t^1 of the first jump and the initial state x^0, and so on. We can choose these conditional distributions such that they satisfy the following conditions:

- The measures $\lambda^k\left(ds \mid x^0, \ldots, x^{k-1}, t^1, \ldots, t^{k-1}; a(\cdot)\right)$ and $\mathsf{P}^k\left(dx^k \mid x^0, \ldots, x^{k-1}, t^1, \ldots, t^k; a(\cdot)\right)$ are measurable with respect to $\mathfrak{X}^k \times \left(\mathbb{B}[0,T]\right)^{k-1} \times \mathfrak{A}^{[0,T]}$ and $\mathfrak{X}^k \times \left(\mathbb{B}[0,T]\right)^k \times \mathfrak{A}^{[0,T]}$ respectively;
- For any Borel-measurable $\Gamma \subset [t, t+h]$, the measure $\lambda^k\left(\Gamma \mid x^0, \ldots, x^{k-1}, t^1, \ldots, t^{k-1}; a(\cdot)\right)$ depends for $t^{k-1} < t$ on $a(\cdot)$ only on $a(s), s \in [0, t+h)$, and $\mathsf{P}^k\left(dx^k \mid x^0, \ldots, x^{k-1}, t^1, \ldots, t^k; a(\cdot)\right)$ depends on $a(\cdot)$ only through its values on $[0, t^k)$.

To construct a representation of the controlled object we need the following auxiliary results.

Lemma 2.28 (see [GS79, Lemma 2.2]). Let X be a complete separable metric space with Borel σ-algebra $\mathfrak{X}$, A be some topological space with Borel σ-algebra $\mathfrak{A}$. Let $\{\mu_a(\cdot), a \in A\}$ be a set of measures over $\mathfrak{X}$ such that $\mu_a(C)$ is $\mathfrak{A}$-measurable for all $C \in \mathfrak{X}$. Then there exists a function

$f(\zeta, a)$ from $[0, 1] \times A$ to X, measurable with respect to $\mathbb{B}[0, 1] \times \mathfrak{A}$. which satisfies the following conditions:

    1) if $\mu_{a^1} = \mu_{a^2}$, then $f(\zeta, a^1) = f(\zeta, a^2)$ for any ζ;

    2) if m_L is the Lebesgue measure on $[0; 1]$, then for all $C \in \mathfrak{X}$ and $a \in A$

$$\mu_a(C) = m_L\left(\left\{\zeta : f(\zeta, a) \in C\right\}\right). \qquad \odot$$

Lemma 2.29 (see [GS79, Lemma 2.3]). Let $(B, \mathfrak{B})$ be a measurable space and $\mu(\cdot \mid b, a(\cdot))$ be a family of distributions on $\mathbb{B}[0, T]$, where $b \in B$ and $a \in A^{[0,T]}$, which satisfies the following conditions:

    1) $\mu(\cdot \mid b, a(\cdot))$ is measurable with respect to $\mathfrak{B} \times \mathfrak{A}^{[0,T]}$;

    2) if $\Gamma \in \mathbb{B}[0, t]$ and $a^1(s) = a^2(s)$ for $s \leq t$, then

$$\mu\left(\Gamma \mid b, a^1(\cdot)\right) = \mu\left(\Gamma \mid b, a^1(\cdot)\right) \quad \text{for any } b.$$

Then there exists a real valued function $\rho(\zeta, b, a(\cdot))$, defined on $[0, T] \times B \times A^{[0,T]}$, measurable with respect to $\mathbb{B}[0, T] \times \mathfrak{B} \times \mathfrak{A}^{[0,T]}$, which possesses the following properties:

    A) $m_L\left(\left\{\zeta : \rho(\zeta, b, a(\cdot)) \in \Gamma\right\}\right) = \mu(\Gamma \mid b, a(\cdot))$ for all Borel sets $\Gamma \in \mathbb{B}[0, T]$;

    B) If $a^1(s) = a^2(s)$ for $s < t$, then $\rho(\zeta, b, a^1(\cdot)) = \rho(\zeta, b, a^2(\cdot))$ for all ζ, for which $\rho(\zeta, b, a^1(\cdot)) \leq t$ holds. $\qquad \odot$

    Utilizing the above lemmas, we now sketch the construction of a controlled object with step function paths governed by a step control with the conditional distributions P^k and λ^k from (2.8). By Lemmas 2.28 and 2.29 we find functions

$$\rho^k\left(\zeta, x^0, \ldots, x^{k-1}, t^1, \ldots, t^{k-1}, a(\cdot)\right) \text{ on } [0, 1] \times X^k \times [0, T]^{k-1} \times A^{[0,T]},$$

$$f^k\left(\zeta, x^0, \ldots, x^{k-1}, t^1, \ldots, t^k, a(\cdot)\right) \text{ on } [0, 1] \times X^k \times [0, T]^k \times A^{[0,T]}$$

with values in $[0, T]$ and X respectively, measurable in all variables, such that the following holds:

1) if Γ is a Borel set in $[0, T]$, then

$$m_L\left(\left\{\varsigma : \rho^k\left(\varsigma, x^0, \ldots, x^{k-1}, t^1, \ldots, t^{k-1}, a(\cdot)\right) \in \Gamma\right\}\right)$$
$$= \lambda^k\left(\Gamma \mid x^0, \ldots, x^{k-1}, t^1, \ldots, t^{k-1}, a(\cdot)\right);$$

and if $C \in \mathfrak{X}$, then

$$m_L\left(\left\{\varsigma : f^k\left(\varsigma, x^0, \ldots, x^{k-1}, t^1, \ldots, t^k, a(\cdot)\right) \in C\right\}\right)$$
$$= \mathsf{P}^k\left(C \mid x^0, \ldots, x^{k-1}, t^1, \ldots, t^k, a(\cdot)\right);$$

2) if $t^1 < \cdots < t^{k-1} < t$, $a^1(s) = a^2(s)$ for $s < t$, then

$$\rho^k\left(\varsigma, x^0, \ldots, x^{k-1}, t^1, \ldots, t^{k-1}, a^1(\cdot)\right)$$
$$= \rho^k\left(\varsigma, x^0, \ldots, x^{k-1}, t^1, \ldots, t^{k-1}, a^2(\cdot)\right)$$

for all ς, for which $\rho^k\left(\varsigma, x^0, \ldots, x^{k-1}, t^1, \ldots, t^{k-1}, a^1(\cdot)\right) \leq t$.

Now let $\varsigma^0, \zeta^1, \varsigma^1, \zeta^2, \varsigma^2, \ldots$ be a sequence of independent uniformly distributed variables on some probability space. Put

$$
\left.
\begin{aligned}
\hat{\xi}^0(\omega) &= f^0\left(\varsigma^0\right), \\
\hat{\sigma}^1\left(\omega, a(\cdot)\right) &= \rho^1\left(\zeta^1, f^0\left(\varsigma^0\right), a(\cdot)\right), \\
\hat{\xi}^1\left(\omega, a(\cdot)\right) &= f^1\left(\varsigma^1, \hat{\xi}^0(\omega), \hat{\sigma}^1\left(\omega, a(\cdot)\right), a(\cdot)\right), \\
&\cdots, \\
\hat{\sigma}^k\left(\omega, a(\cdot)\right) &= \rho^k\left(\zeta^k, \hat{\xi}^0(\omega), \ldots, \hat{\xi}^k\left(\omega, a(\cdot)\right),\right. \\
&\qquad \left. \hat{\sigma}^1\left(\omega, a(\cdot)\right), \ldots, \hat{\sigma}^{k-1}\left(\omega, a(\cdot)\right), a(\cdot)\right), \\
\hat{\xi}^k\left(\omega, a(\cdot)\right) &= f^k\left(\varsigma^k, \hat{\xi}^0(\omega), \ldots, \hat{\xi}^{k-1}\left(\omega, a(\cdot)\right),\right. \\
&\qquad \left. \hat{\sigma}^1\left(\omega, a(\cdot)\right), \ldots, \hat{\sigma}^k\left(\omega, a(\cdot)\right), a(\cdot)\right), \\
&\cdots
\end{aligned}
\right\}
$$

From 1) and 2) we conclude:

If $a^1(s) = a^2(s)$ for all $s < t$, then for all ω with $\hat{\sigma}^k(\omega, a^1(\cdot)) \leq t$ we have

$$\hat{\sigma}^k(\omega, a^1(\cdot)) = \hat{\sigma}^k(\omega, a^2(\cdot)), \text{ and } \hat{\xi}^k(\omega, a^1(\cdot)) = \hat{\xi}^k(\omega, a^2(\cdot)).$$

Now, the joint distribution of the so constructed random variables $\hat{\xi}^0(\omega)$, $\hat{\sigma}^1(\omega, a(\cdot)), \ldots, \hat{\sigma}^k(\omega, a(\cdot))$ and $\hat{\xi}^k(\omega, a(\cdot))$ coincides with the joint distribution of the values $\xi^0, \sigma^1, \ldots, \sigma^k, \xi(\sigma^k)$ under the probability measure $\mu(\cdot \mid a(\cdot))$.

Therefore, with $\sigma^0 = 0$ the random function ξ defined by

$$\xi(t, \omega, a(\cdot)) = \hat{\xi}^k(\omega, a(\cdot)), \quad \text{for all } t \in \left[\hat{\sigma}^{k-1}(\omega, a(\cdot)), \hat{\sigma}^k(\omega, a(\cdot))\right)$$

is a representation of the controlled object we wanted to find.

2.3.2 Markov jump processes with step control

In this subsection we consider processes with time scale $\mathbb{R}_+$ and specialize first the general definition of control and controlled object to the Markovian setting. We consider only Markov processes with Polish state space X and compact action space A. $\mathfrak{X}$ and $\mathfrak{A}$ are the respective Borel-σ-algebras.

Definition 2.30. A controlled Markov process with Polish state space $(X, \mathfrak{X})$ of the basic process and compact action space $(A, \mathfrak{A})$ for the control is defined by a set of consistent transition probabilities $P(t, x, s, C; a(\cdot))$ for $0 \leq t < s < \infty$, $x \in X$, $C \in \mathfrak{X}$, $a(\cdot) \in A^{[0,\infty)}$. The transition probabilities are measurable for fixed $t < s$ and C with respect to $\mathfrak{X} \times \mathfrak{A}^{[t,s)}$. From the (controlled) transition probabilities $P(t, x, s, C; a(\cdot))$ we can derive a family $\mu_x(\cdot \mid \cdot)$ of Markov process distributions, which depend on the initial state of the basic process x as a parameter. For every $a(\cdot) \in A^{[0,\infty)}$ the family of measures $\mu_x(\cdot \mid a(\cdot))$ corresponds to a Markov process with transition probability $P(\cdot, \cdot, \cdot; a(\cdot))$. ⊙

Remark 2.31. If the controls $a(\cdot)$ are step functions then for determining the process distributions, it is sufficient to know transition probabilities of the form $\mathsf{P}\big(t, x, s, C; a(\cdot)\big) = \mathsf{P}(t, x, s, C; a)$, where the controls are constant functions $a(\cdot) \equiv a$. From these transition functions we can construct the transition functions under general step control with the aid of the Markov property as follows:

Assume that for a general step control $a(\cdot)$ with jump times $0 = t^1 < t^2 < \cdots < t^n < \ldots$ we have $a(t) = a^k$ for $t^k \le t < t^{k+1}$ then for $t < s$ with $t^{j-1} \le t < t^j < \cdots < t^n \le s < t^{n+1}$ we have

$$
\mathsf{P}\big(t, x, s, C; a(\cdot)\big) = \int \mathsf{P}\big(t, x, t^j, dx^j; a^{j-1}\big) \times
$$
$$
\times \int \mathsf{P}\big(t^j, x^j, t^{j+1}, dx^{j+1}; a^j\big) \cdots \int \mathsf{P}\big(t^n, x^n, s, dx^{n+1}; a^n\big). \quad \odot
$$

Our main interest is in the class of controlled Markov jump processes under step control. These processes are connected with step controls and controlled objects $\mu(\cdot \mid a(\cdot))$, which are concentrated on the space of step functions and pose a Markov property for a given control. We further require that the inter-jump times $\sigma^{k+1} - \sigma^k$, $k \in \mathbb{N}$, have bounded densities.

Definition 2.32. A controlled Markov jump process is specified by the following properties:

Let the state and action space be as in Definition 2.30 and denote by $D\big([0, \infty), A\big)$ the set of functions on $[0, \infty)$ with values in A being right continuous with left-hand limits.

For the family of transition functions $\mathsf{P}\big(t, x, s, C; a(\cdot)\big)$ from Definition 2.30 for all $0 \le t < s < \infty$ and $x \in X$, $C \in \mathfrak{X}$ with $a(\cdot) \in \bar{A}^{[0,\infty)} \cap D\big([0, \infty), A\big)$ (the space of right continuous step functions without discontinuities of the second kind) the right derivative

$$
\lim_{s \downarrow t} \frac{1}{s - t} \Big[\mathsf{P}\big(t, x, s, C; a(\cdot)\big) - \mathbb{1}_C(x) \Big] = \Pi\big(t, x, a(\cdot), C\big),
$$

exists, and the limit function $\Pi(t, x, a, C)$ is continuous in t, jointly measurable in t, x, a, σ-additive in C, and the function

$$\Pi\big(t, x, a, \{x\}\big) := -\Pi\big(t, x, a, X - \{x\}\big)$$

is bounded. $\odot$

If the control of the Markov jump process is not a deterministic function $a(\cdot)$ as suggested in the above definition, the construction of the controlled process needs some care. For a given randomized step control $\nu(\cdot \mid x(\cdot))$, which governs the development of the process, we now sketch the time development of a controlled Markov jump process similarly to the construction on page 31. We define a sequence of processes $\Big\{\big(\xi_n(t), \alpha_n(t) : t \geq 0\big) : n \in \mathbb{N}\Big\}$ as follows:

If $\xi(0) = x^0$ is the initial state of the basic process, then we define $x_0(t) = x^0$, $0 \leq t \leq \infty$, and $\xi_0 = x_0$. The control process $\alpha_0(t)$ is then governed by the distribution $\nu(\cdot \mid x_0(\cdot))$.

Let $\xi_1(t)$ be a jump process, for which $\xi_1(0) = x^0$, and the time of the first jump σ^1 has the conditional distribution

$$\Pr\{\sigma^1 > t \mid \alpha_0 = a^0\} = \exp\left\{\int_0^t \Pi\big(s, x^0, a^0(s), \{x^0\}\big) ds\right\}.$$

We prescribe

$$\Pr\Big\{\xi_1\big(\sigma^1 + 0\big) \in C \mid \sigma^1 = t, \alpha_0 = a^0\Big\} = \frac{\Pi\big(t, x^0, a_0(t), C - \{x^0\}\big)}{\Pi\big(t, x^0, a_0(t), X - \{x^0\}\big)},$$

and if $\xi_1\big(\sigma^1 + 0\big) = x^1$, we set then $\xi_1(t) = \xi_1\big(\sigma^1 + 0\big) = x^1$ for $t > \sigma^1$. The associated control is the process $\alpha_1(\cdot)$, which coincides with $\alpha_0(\cdot)$ until time σ^1, and then develops such that the conditional distribution of $\alpha_1(\cdot)$ given $x^0, \sigma^1, x^1, \alpha_1(\cdot)$ coincides with that of $\nu(\cdot \mid \xi_1(\cdot))$.

Define the conditional distribution of the time σ^2 of the second jump conditioned on $x^0, \sigma^1, x^1, a^1(\cdot)$ by

$$\Pr\left\{\sigma^2 > t \mid a^1(\cdot), x^0, x^1, \sigma^1 = s\right\}$$

$$= \begin{cases} 1, & t \leq s; \\ \exp\left\{\displaystyle\int_s^t \Pi\left(r, x^1, a^1(r), \{x^1\}\right) dr\right\}, & t > s, \end{cases}$$

and give $\xi_2(\sigma^2 + 0)$ the following conditional distribution:

$$\Pr\left\{\xi_2(\sigma^2 + 0) \in C \mid a^1(\cdot), x^0, x^1, \sigma^1 = s, \sigma^2 = t\right\}$$

$$= \frac{\Pi\left(t, x^1, a^1(t), C - \{x^1\}\right)}{\Pi\left(t, x^1, a^1(t), X - \{x^1\}\right)}.$$

Continuing this way, we see that we have constructed a sequence of processes $\left((\xi_n(t), \alpha_n(t)): t \geq 0\right)$ for $n = 2, 3, \ldots$, with joint jump times σ^k, which satisfy the following conditions:

1) $\xi_n(t)$ is a right continuous step process, which has exactly n jumps, which are $\sigma^1, \ldots, \sigma^n$;

2) $\xi_{n-1}(t) = \xi_n(t)$ for $t < \sigma^n$;

3) $\alpha_{n-1}(t) = \alpha_n(t)$ for $t < \sigma^n$;

4) let $\{\mathcal{F}_t^n, t \geq 0\}$ be the natural filtration of the process $\alpha_n = (\alpha_n(t): t \geq 0)$, $\mathcal{F}_\infty^n = \sigma\left(\bigcup_{t \geq 0} \mathcal{F}_t^n\right) = \sigma(\alpha(t): t \geq 0)$ and $\mathfrak{M}^n = \sigma(\xi(t): t \geq 0)$ the σ-algebras generated by the processes α_n and ξ_n respectively. Then

$$\Pr\left\{\sigma^{n+1} > t \mid \mathcal{F}_\infty^n, \mathfrak{M}^n\right\}$$

$$= \exp\left\{\int_{\sigma^n}^{t \vee \sigma^n} \Pi\left(s, \xi_n(\sigma^n), \alpha_n(s), \{\xi_n(\sigma^n)\}\right) ds\right\}, \quad (2.9)$$

$$\Pr\left\{\xi(\sigma^{n+1}) \in C \mid \mathcal{F}_\infty^n, \mathfrak{M}^n, \sigma^{n+1}\right\}$$

$$
= \frac{\Pi\left(\sigma^{n+1}, \xi_n(\sigma^n), \alpha_n(\sigma^{n+1}), C - \left\{\xi_n(\sigma^n)\right\}\right)}{\Pi\left(\sigma^{n+1}, \xi_n(\sigma^n), \alpha_n(\sigma^{n+1}), X - \left\{\xi_n(\sigma^n)\right\}\right)}, \tag{2.10}
$$

where $\Pr\{\cdot \mid \mathcal{F}_\infty^n, \mathfrak{M}^n\}$ is the conditional probability with respect to σ-algebra generated by $\mathcal{F}_\infty^n$, $\mathfrak{M}^n$; in the second case, the conditioning σ-algebra is generated by $\mathcal{F}_\infty^n$, $\mathfrak{M}^n$ and σ^{n+1};

5) let $\mathfrak{N}^n$ be a σ-algebra generated by the events of the following type:

$$
\{\sigma^n > t\} \cap C^t \cap D,
$$

where $C^t \in \mathcal{F}_t^n$, $D \in \mathfrak{M}^n$, $t > 0$; then for all sets B from $\mathfrak{A}^{[t,\infty]}$ we have

$$
\Pr\{\alpha_{n+1}(\cdot) \in B \mid \mathfrak{N}^n\} = \nu\left(B \mid \xi_n(\cdot)\right) \tag{2.11}
$$

for $\sigma^n \leq t$.

One can verify that (2.9)–(2.11) and the conditions 1, 2, 3 uniquely determine the joint distributions of the processes $\alpha_n(t)$ and $\xi_n(t)$, if only $\xi_0(0) = \xi_n(0)$ is given. Put $\xi(t) = \xi_n(t)$, $\alpha(t) = \alpha_n(t)$, $t \in [\sigma^n; \sigma^{n+1})$, where $\sigma^0 = 0$. Note that from the boundedness of the inter-jump time intensities we have $\Pr\{\sigma^n \to \infty\} = 1$. Then the processes $\xi(t)$ and $\alpha(t)$ are defined on $[0; \infty)$. The pair of processes $\xi = (\xi(t) : t \geq 0)$ and $\alpha = (\alpha(t) : t \geq 0)$ constructed this way is a controlled Markov jump process with the given controlled object and control ν.

If $\nu(\cdot \mid \cdot)$ is of the following form:

For $t > 0$, $B \in \mathcal{A}^{[t,t]}$, and $x(\cdot)$ given, we have $\nu\left(B \mid x(\cdot)\right) = \mathbb{1}_B\left(\eta(t, x)\right)$, where $\eta(t, x)$ is a deterministic function, then the construction is much simpler. (We have a *non-randomized Markov control*.) From (2.9) and (2.10) it follows, that $\xi(t)$ is a Markov jump process with transition probability $\mathsf{P}_\eta(t, x, s, C)$ satisfying

$$
\lim_{s \downarrow t} \frac{1}{s - t} \left[\mathsf{P}_\eta(t, x, s, C) - \mathbb{1}_C(x)\right] = \Pi\left(t, x, \eta(t, x), C\right).
$$

2.4   TOPOLOGICAL FOUNDATIONS

Before we present the details of our stochastic optimization problems, we recall some Definitions and Theorems from [Kur69] and [HV69]. These are definitions of multivalued functions (set valued functions) and theorems on the existence of smooth functions that select from the set valued image of such functions a single value. The theorems are therefore know as theorems on the existence of smooth selectors.

Definition 2.33. Given nonempty sets X and A; a multivalued function (multifunction) $F\colon X \to A$ is a function on X such that each value $F(x)$ is a nonempty subset of A.

   If we denote by 2^A the set of all subsets of A, then a multifunction is a function

$$F\colon X \to 2^A - \{\emptyset\},$$

i.e., a set valued function with domain X.

   If $B \subseteq A$, then

$$F^{-1}(B) := \{x \in X : F(x) \cap B \neq \emptyset\}.$$

   A function $f\colon X \to A$ is a selector for the multifunction F if $f(x) \in F(x)$ for all $x \in X$. $\odot$

   Typical examples of multifunctions (set valued functions) are the elements of the sequences $A = \left(A^t \colon t \in \mathbb{N}\right)$ from Definition 2.16 that determine the admissible actions in a decision model.

Definition 2.34. For topological spaces X and A with Borel σ-algebras $\mathfrak{X}$ and $\mathfrak{A}$ a map $F\colon X \to 2^A - \{\emptyset\}$, and the associated multifunction $F\colon X \to A$ are point-closed if for all $x \in X$ the subset $F(x) \subseteq A$ is closed.

   A point-closed map F is

- open-measurable, if for all open sets $E \subseteq A$ we have $F^{-1}(E) \in \mathfrak{X}$,

- closed-measurable, if for all closed sets $E \subseteq A$ we have $F^{-1}(E) \in \mathfrak{X}$,

and

- Borel-measurable, if for all Borel sets $E \subseteq A$ we have $F^{-1}(E) \in \mathfrak{X}$.

A point-closed map F is upper semicontinuous, if for all closed sets $E \subseteq A$ the set $F^{-1}(E)$ is closed.

A point-closed map F is lower semicontinuous, if for all open sets $E \subseteq A$ the set $F^{-1}(E)$ is open.

A mapping F is continuous if it is simultaneously upper and lower semi-continuous. ⊙

Theorem 2.35 (Selection theorem; see [Kur69, p. 74], [KRN65, AL72]). Let $(X, \mathfrak{X})$ be a measurable space, and let A be a complete separable metric space. If a point-closed map is according to Definition 2.34 closed-, open-, or Borel-measurable, then it has a Borel measurable selector.⊙

Theorem 2.36 (Selection theorem for semicontinuous maps [Kur69, p. 74]). Let $(X, \mathfrak{X})$ be a measurable space, and let A be a complete separable metric space. Then any semicontinuous map $F \colon X \to 2^A - \{\emptyset\}$ has a selector belonging to Baire class 1. ⊙

Corollary 2.37. Let $(X, \mathfrak{X})$ be a measurable space, and let A be a compact space with countable basis. Then any semicontinuous map $F \colon X \to 2^A - \{\emptyset\}$ has a selector belonging to Baire class 1. ⊙

Chapter 3

LOCAL CONTROL OF DISCRETE TIME INTERACTING MARKOV PROCESSES WITH GRAPH STRUCTURED STATE SPACE

In this chapter we develop models for spatially distributed systems that are controlled by decision makers who act locally and have only information at hand about the system's state in the neighborhood around their position. To optimize such systems poses problems due to the restrictions on the information gain that arise from only observing locally the system over time. Another source of difficulties may be a cost function that reacts on the global behavior of the system, although only local control is possible for the decision makers.

We consider in this chapter only systems that are observed at equally spaced time instants; the time parameter of the underlying stochastic processes will be $\mathbb{N}$. Discrete time systems, especially queueing models, have found increasing interest in recent years due to the occurrence of models with a generic inherent discrete time scale.

Such control problems with specific underlying neighborhood systems occur, e.g., in modeling transmission systems, communication networks, production systems, and distributed populations. For systems with states that incorporate such a neighborhood structure, we construct system processes governed by controls that are compatible with these neighborhood structures, and study their evolution over time. The general class of *local* policies obtained in this way gives rise to problems and classifications analogously to those in the classical theory of stochastic dynamic optimization. An important question that arises is whether we may restrict our search for optimal policies from the very beginning to smaller classes of special *local* policies, say, e.g., *stationary Markovian* policies. This property is well known to hold in many situations where decisions are made on the basis of global information.

The standard stochastic models for spatially distributed systems are *random fields* over some multidimensional lattice $\mathbb{Z}^d$, or subsets thereof, or over some general graph Γ. Control of stochastically driven systems over time is part of the theory of stochastic processes. Therefore our generic model in this chapter is a spatially distributed process evolving in time. The fundamental theory especially connected with the origins of those models in statistical physics can be found in [Lig85]. The class of stochastic processes used there is that of Markovian processes in time, and for the spatial structure: *Gibbs measures,* which are essentially equivalent to *Markov random fields.*

It is well known that in continuous time, Markov processes with local transition structure over some (nonregular) graph may not have Markov random fields with respect to that graph as one-dimensional marginals nor as limiting random fields; see [Kel79, Section 9]. Similar problems arise in our discrete time setting. We describe a class of processes throughout this chapter which, to a certain extent, resemble Kelly's *Spatial Processes* ([Kel79, Section 9, page 189]; see as well [Whi86]). But we do not impose his restriction that state changes are of strong local character, such that only one coordinate may change its state at a time instant. We allow several coordinates of a state may change simultaneously, which is essential for discrete time systems, where *simultaneous events* usually occur.

3.1 RANDOM FIELDS WITH GRAPH STRUCTURED STATE SPACES OF PRODUCT FORM

The mathematical theory of random fields that was developed over the last fifty years provides us with standard models for the investigation of high dimensional complex systems in the areas of technical, physical, biological, and economical research.

From a mathematical point of view, the first problem was to describe consistently random field models. Many results obtained in statistical physics for Gibbs measures have been fruitfully generalized to other areas. References to the field are, e.g., [Dob68, Dob71, BGM69, Lig85, Sin80, MM85, Rue69, Sul75, Ave72, VK78]. A comprehensive introduction into the mathematics of random fields is [Geo88].

Gibbs states are models to describe structured interaction in systems with a large number of components in equilibrium. Assuming the system to be in equilibrium the random law of the state does not change with time. In this way we next define the random law of our systems at a fixed time instant as a random field over a graph. The graph describes the interaction structure of the systems.

For systems with locally interacting coordinates, the interaction structure is defined via an undirected finite neighborhood graph $\Gamma = (V, B)$ without loops and double edges. The graph has set of vertices (nodes) V and set of edges B. Denote by $\{k, j\}$ the edge of the graph connecting vertices k and j. The neighborhood of vertex k is the node set $N(k) = \{j : \{k, j\} \in B\}$. The complete neighborhood of vertex k is $\widetilde{N}(k) = N(k) \cup \{k\}$; i.e., the neighborhood of the vertex k, and including k. For any $K \subset V$ we define the neighborhood $N(K) = \bigcup_{k \in K} N(k) - K$, and the complete neighborhood $\widetilde{N}(K) = N(K) \cup K$.

Definition 3.1. A nonempty subset $C \subseteq V$ of a graph $\Gamma = (V, B)$ is a clique or simplex in Γ if either $|C| = 1$ or all nodes of C are neighbors, i.e.,

C is a complete subgraph of Γ. The set of all cliques is $\mathcal{C}$. $C \in \mathcal{C}$ is a maximal clique if for all $i \in V - C$ holds $C \cup \{i\} \notin \mathcal{C}$.   ⊙

For every node $i \in V$ let $(X_i, \mathfrak{X}_i)$, $X_i \neq \emptyset$, be some Polish measurable space equipped with the Borel-σ-algebra $\mathfrak{X}_i$. X_i, respectively $(X_i, \mathfrak{X}_i)$ is called the local state space at node i. We further equip $X := \underset{i \in V}{\times} X_i$ with the product σ-algebra $\mathfrak{X} = \sigma \left\{ \underset{i \in V}{\times} \mathfrak{X}_i \right\}$ generated by $\underset{i \in V}{\times} \mathfrak{X}_i$. X, respectively $(X, \mathfrak{X})$, is the global state space of the system. For every subset of nodes $K \subset V$ we denote the marginal vector of state $x = (x_i : i \in V)$ by $x_K = (x_k : k \in K) \in X_K = \underset{i \in K}{\times} X_i$.

We define for $K \subseteq V$ the σ-algebra $\mathfrak{X}_K = \sigma \left\{ \underset{i \in K}{\times} \mathfrak{X}_i \right\} = \underset{i \in K}{\bigotimes} \mathfrak{X}_i$, generated by $\underset{i \in K}{\times} \mathfrak{X}_i$, so $\mathfrak{X}_V = \mathfrak{X}$.

Definition 3.2 (Random fields). (1) A random variable

$$\xi \colon (\Omega, \mathcal{F}, \mathrm{Pr}) \to (X, \mathfrak{X}) = \left(\underset{i \in V}{\times} X_i, \underset{i \in V}{\bigotimes} \mathfrak{X}_i \right)$$

is called a random field over $\Gamma = (V, B)$ (or simply a random field over V). For $\emptyset \neq K \subseteq V$, the marginal random variables with values in the space X_K are denoted by ξ_K. For $K = \{k\}$ we write ξ_k.

(2) A random field ξ on (V, B) is a Markov field if for all $k \in V$ the following holds:

$$\mathrm{Pr} \left\{ \xi_k \in C_k \mid \xi_{V-\{k\}} = x_{V-\{k\}} \right\} = \mathrm{Pr} \left\{ \xi_k \in C_k \mid \xi_{N(k)} = x_{N(k)} \right\},$$

$$\forall\, x \in X, \quad C_k \in \mathfrak{X}_k, \quad (3.1)$$

where $\mathrm{Pr} \left\{ \xi_k \in C_k \mid \xi_{N(k)} = x_{N(k)} \right\}$, respectively $\mathrm{Pr} \left\{ \xi_k \in C_k \mid \xi_{V-\{k\}} = x_{V-\{k\}} \right\}$ is the regular conditional probability on $(X_k, \mathfrak{X}_k)$ given $\xi_{N(k)} = x_{N(k)}$, respectively $\xi_{V-\{k\}} = x_{V-\{k\}}$.

(3) (Canonical random field) We often assume that $(\Omega, \mathcal{F}, \mathrm{Pr})$ in the definition part **(1)** is the canonical space. Then ξ_K and ξ_k are the respective projections and we have $\mathrm{Pr}^\xi = \mathrm{Pr}$.

(4) (Discrete states) If all the local state spaces are discrete then we usually identify the distribution of the random field and its counting density and write in case of having the canonical underlying probability space $\mathrm{Pr}(\xi = x) = \mathrm{Pr}^\xi\{x\} = \mathrm{Pr}(x)$, $x \in X$. ⊙

Remark 3.3. Equation (3.1) reads as follows:

$$\mathrm{Pr}\left\{\xi_k \in C_k \mid \xi_{V-\{k\}} = x_{V-\{k\}}\right\} = \mathrm{Pr}\left\{\xi_k \in C_k \mid \xi_{V-\{k\}} = y_{V-\{k\}}\right\}$$
$$\text{if } x_{N(k)} = y_{N(k)} \text{ holds.}$$

Because the conditional probabilities on both sides of the equation are defined only $\mathrm{Pr}^{\xi_{V-\{k\}}}$-almost surely, the required equality is assumed to hold outside of some set with $\mathrm{Pr}^{\xi_{V-\{k\}}}$-probability 0.

For random fields with a discrete state space to be a Markov field, it is often required that the counting density is strictly positive [Kel79, p. 186]. We do not put this into the definition, following [KV80] and others. ⊙

The following lemma ensures that the definition of a Markov field used in [KV80, Definition 7] coincides with our definition. This is of relevance for the Theorems 3.14 and 3.15.

Lemma 3.4. A strictly positive random filed with discrete state space

$$\xi \colon (\Omega, \mathcal{F}, \mathrm{Pr}) \to (X, \mathfrak{X}) = \left(\underset{i \in V}{\times} X_i, \bigotimes_{i \in V} \mathfrak{X}_i\right)$$

is a Markov field if and only if for all nonempty $K, J \subseteq V$ such that $\widetilde{N}(K) \subseteq J$ holds

$$\Pr\left\{ \xi_K = y_K \mid \xi_{J-\{K\}} = x_{J-\{K\}} \right\} = \Pr\left\{ \xi_K = y_K \mid \xi_{N(K)} = x_{N(K)} \right\}$$

$$\forall x \in X, \quad y_K \in X_K. \quad \odot$$

Example 3.5. Suppose we are given a discrete state random field P according to Definition 3.2 over $\Gamma = (V, B)$.

(a) Assume that $V \in \mathcal{C}$ is a simplex, i.e., Γ is a complete graph with $B = \{\{i, j\} \subseteq V : i < j\}$. Then for all $j \in V$: $N(j) = V - \{j\}$. So P is a Markov random field.

(b) Assume that $B = \emptyset$, i.e., $\Gamma = (V, B)$ totally disconnected. Then for all $j \in V$: $N(j) = \emptyset$. So P is a Markov field if and only if it has independent coordinates:

$$P(x_j \mid x_{V-j}) = P(x_j) = \Pr(\xi_j = x_j).$$

(c) If the coordinates of P are independent as in (b), then P is Markov field with any underlying graph Γ. $\qquad\qquad\qquad\qquad\qquad \odot$

An attempting way to construct random fields is to specify the local stochastic behavior at the nodes via the conditional distribution at this node given the states of the other nodes are prescribed (frozen).

Definition 3.6 (Specification). Suppose we are given a discrete state random field $\xi \colon (\Omega, \mathcal{F}, \Pr) \to (X, \mathfrak{X})$ over $\Gamma = (V, B)$ according to Definition 3.2.

(a) For $x = (x_j : j \in V) \in X$ and $j_0 \in V$ define

$$P\left(x_{j_0} \mid x_{V-j_0}\right) := \Pr\left(\xi_{j_0} = x_{j_0} \mid \xi_{V-j_0} = x_{V-j_0}\right)$$

$$= \frac{\Pr(\xi = x)}{\sum_{y \in X_{j_0}} \Pr\left(\xi_{j_0} = y, \xi_{V-j_0} = x_{V-j_0}\right)}$$

if the denominator is greater than 0.

If the denominator is 0, we set $\Pr\left(x_{j_0} \mid x_{V-j_0}\right) = 0$.

This set of conditional distributions is called a *specification*.

(b) For $H, G \subseteq V$, $H, G \neq \emptyset$, $x = (x_j : j \in V) \in X$ we set

$$P(x_H \mid x_G) := \begin{cases} \Pr(\xi_H = x_H \mid \xi_G = x_G), & \text{if } \Pr(\xi_G = x_G) > 0; \\ 0, & \text{elsewhere.} \end{cases} \qquad \odot$$

Remark 3.7. Identification and construction of admissible specifications is not an easy task. The book of Georgii [Geo88] is an in-depth study of this question and whether a specification defines uniquely a distribution of a random field. A more recent study concerning modeling problems related to this question is [KC00]. $\qquad \odot$

A closely related class of distributions for random fields are well known in statistical physics where the most interesting questions arise with infinite graphs. We describe here the case of discrete state spaces and finite graphs along the lines of [KV80].

Definition 3.8 (Potential). Let $\Gamma = (V, B)$ be an interaction graph and $X := \underset{i \in V}{\times} X_i$ a discrete graph structured state space.

(a) A *potential* (or *interaction potential*) on (Γ, X) is a collection $u = (u_C : C \subseteq V)$ of functions $u_C \colon \underset{i \in C}{\times} X_i \to \mathbb{R}$.

A potential $u = (u_C : C \subseteq V)$ is called *Gibbs potential* if u_C vanishes (is constant zero) for all $C \subseteq V$ that are not a clique.

(b) For a given (Gibbs) potential u and any subgraph $J \in V$, the *energy* $U(x_J)$ of the (local) configuration $x_J \in \underset{i \in J}{\times} X_i$ is

$$U(x_J) = \sum_{\substack{C \subseteq J \\ C \in \mathcal{C}}} u_C(x_C).$$

(c) For $K \subset J$, $x_J \in \underset{i \in J}{\times} X_i$ we define the "conditional energy" as difference of the respective energies

$$U\big(x_K \mid x_{J \setminus K}\big) := U(x_J) - U(x_K). \qquad \odot$$

We can define Gibbs fields by their specifications.

Definition 3.9 (Gibbs field). Let $\Gamma = (V, B)$ be an interaction graph and $\xi \colon (\Omega, \mathcal{F}, \mathrm{Pr}) \to (X, \mathfrak{X})$ a random field with graph structured state space $X := \underset{i \in V}{\times} X_i$ and let $u_C \colon \underset{i \in C}{\times} X_i \to \mathbb{R}$ be a Gibbs potential. Then ξ is a *Gibbs field* on (Γ, X) with Gibbs potential u if for all $K, J \subseteq V$, with $N(K) \subseteq J$, $x_J \in \underset{i \in J}{\times} X_i$ we have

$$
\begin{aligned}
\mathsf{P}\big(x_K \mid x_{J \setminus K}\big) &= \mathrm{Pr}\left(\xi_K = x_K \mid \xi_{J \setminus K} = x_{J \setminus K}\right) \\
&= \frac{\exp\left[-U\big(x_K \mid x_{J \setminus K}\big)\right]}{\sum_{z_K \in X_K} \exp\left[-U\big(z_K \mid x_{J \setminus K}\big)\right]}.
\end{aligned} \qquad (3.2)
$$

$\odot$

Note that the interaction graph for the Gibbs field comes into the above definition via the clique structure of Γ, which is used in the definition of the Gibbs potential.

It has been proven that any strictly positive Markov random field in the sense of Definition 3.2 is a Gibbs field and vice versa; see [Ave72, Gri73].

Random fields can be looked upon as stochastic processes with *nonlinear time scale V*. The lack of linearity of the time scale gave rise to intense research; see the references indicated above. There is a well developed theory now even for infinite regular underlying lattice graphs, say $\mathbb{Z}^n$, with nearest neighborhood structures. An important problem is to define an analogy to the Markov property in time, which makes a random field to be a Markov random field. Several definitions were suggested in the literature, and it turned out that the most important could be proven to be equivalent.

3.2 INTERACTING MARKOV PROCESSES IN DISCRETE TIME

In many applications in biology, in economics, or engineering research, random fields describe the state of a system for some fixed time instant. The evolution over time of such systems is then described by a stochastic process η with random fields as one-dimensional marginals in time. In this case we write $\xi = (\xi^t)$ to emphasize the time-dependence.

We assume that t takes discrete values $t = 0, 1, \ldots$. The subscript k in ξ_k^t refers to the vertex k, ξ_k^t therefore denotes the marginal variable for time t and node k of some vector valued process $\xi = (\xi^t : t = 0, 1, \ldots)$ with state space $(X, \mathfrak{X})$. In this chapter we consider only Markov processes in discrete time.

If a Markov process ξ from its very construction is related to the neighborhood system $\{N(k) : k \in V\}$ of (V, B), it is natural to assume that with respect to the evolution over time, the probability for the event $\{\tilde{\xi}_k^t = x_k\}$ at the k-th vertex depends on the previous states of the whole system only through the values of the vertices in $\tilde{N}(k)$ (including k) at $t - 1$. To describe this, we introduce the notion of the Markov property in space and in time along the lines of Definition 3.2.

Definition 3.10 (Local and synchronous transition probabilities). Let $\xi = \{\xi^t, t = 0, 1, \ldots\}$, $\xi^t \colon (\Omega, \mathcal{F}, \mathrm{Pr}) \to (X, \mathfrak{X})$, be a discrete time Markov process with state space $(X, \mathfrak{X}) = \left(\underset{i \in V}{\times} X_i, \underset{i \in V}{\bigotimes} \mathfrak{X}_i \right)$. The transition probabilities of ξ are said to be *local* [Vas78, p. 100] if

$$\Pr\left\{\xi_k^{t+1} \in C_k \mid \xi^t = x^t, \dots, \xi^0 = x^0\right\}$$
$$= \Pr\left\{\xi_k^{t+1} \in C_k \mid \xi_{\widetilde{N}(k)}^t = x_{\widetilde{N}(k)}^t\right\}$$
$$\text{for } k \in V, \ x^0, \dots, x^t \in X, \ C_k \in \mathfrak{X}_k, \quad (3.3)$$

holds $\Pr^{(\xi^t, \dots, \xi^0)}$-almost surely; i.e., the transition probability at vertex k depends on the state of its complete neighborhood at the previous time instant only.

The transition probabilities of ξ are said to be *synchronous* [Vas78, p. 100] if

$$\Pr\left\{\xi_K^{t+1} \in C_K \mid \xi^t = x^t\right\} = \prod_{k \in K} \Pr\left\{\xi_k^{t+1} \in C_k \mid \xi^t = x^t\right\},$$
$$\forall \ K \subset V, \ x^t \in X, \ C_K = \underset{k \in K}{\times} C_k \in \mathfrak{X}_K, \quad (3.4)$$

holds $\Pr^{\xi^t}$-almost surely.

If ξ fulfills (3.3) and (3.4), it is called a *Markov process with locally interacting synchronous components over* (Γ, X), shortly, a *(time dependent) Markov random field*. $\qquad \odot$

Some comments on Markov processes with product state spaces and with local and synchronous transition probabilities may be in order here. Consider $\xi = \left(\xi_k^t \colon (t, k) \in \mathbb{N} \times V\right)$ as a general random element:

The first part of the definition ensures that the one-step transition behavior of the time dependent Markov random field in t-direction on some subgraph K of Γ given fixed values of the random element in the neighborhood $N(K)$ in k-direction does not depend on the behavior of the field outside $\widetilde{N}(K)$. The second property guarantees conditional independence of the nodes of the graph evolving in the time direction given the spatial neighborhood in addition to the memoryless property of the nodes with respect to the past given the present local state.

From the very definition, the Markov process $\xi = \{\xi^t, t = 0, 1, \dots\}$ has one-dimensional marginals that are random fields, but this notion does not impose any restrictions in space direction. The requirements of Definition 3.10 will have many appealing interpretations in the context of economics, physics, and biological phenomena.

A typical situation is the following: Let the system consist of a finite or infinite number of particles, the evolution of which is described by independent Markov chains. Then on the evolution of this system some possibly state dependent random or deterministic constraints are imposed. So the evolution of each particle is no longer Markovian, while the evolution of the whole system is Markovian but very complicated. If the imposed constraints are of local character, then Definition 3.10 should apply, due to the original independence.

Remark 3.11. Equations (3.4) and (3.3) localize and synchronize the behavior of the Markov process by requirements for the conditional probabilities that do exist from the assumption of having a Polish state space X and are almost surely valid.

If we are given a transition kernel of the Markov process, then we alternatively can formulate the respective properties by requirements for the transition function. This will be a deterministic relation for that function. If no ambiguity emerges, we shall use both versions alternatively without further comments.   ⊙

If we are given a Markov chain with a transition kernel that is local and synchronous from Definition 3.10 with respect to some graph $\Gamma = (V, B)$, then in general, the one-dimensional distributions of that chain are not Markov random fields over $\Gamma = (V, B)$. The reason is that in general, dependencies are propagated with time evolution. Consequently, the stationary and limiting distributions (if they exist) are in general not a Markov field over $\Gamma = (V, B)$. We comment on this problem, providing some further details and theorems. We restrict ourselves to the case of discrete state space X (finite or countably infinite).

Let $\xi = \{\xi^t, t = 0, 1, \dots\}$, $\xi^t \colon (\Omega, \mathcal{F}, \mathrm{Pr}) \rightarrow (X, \mathfrak{X})$, be a discrete time Markov process with discrete state space $X := \underset{i \in V}{\times} X_i$. Assume that the transition probabilities of ξ are local and synchronous; see Definition 3.10. Then the transition kernel Q of ξ is

$$
\begin{aligned}
\mathsf{Q}(x \mid y) &= \mathrm{Pr}\left\{\xi^{t+1} = x \mid \xi^t = y\right\} \\
&= \prod_{k \in V} \mathrm{Pr}\left\{\xi_k^{t+1} = x_k \mid \xi_{\widetilde{N}(k)}^t = y_{\widetilde{N}(k)}\right\}.
\end{aligned}
\tag{3.5}
$$

We assume in the following that ξ is irreducible on X and denote the steady state distribution of ξ, if it exists, by π. For example, π is the probability solution of

$$
\sum_{x \in X} \mathsf{Q}(y \mid x)\pi(x) = \pi(y), \quad y \in X.
$$

We now define a class of transition kernels that resemble the definition of the specification of Gibbs fields in (3.2). The feasible potentials are restricted as follows.

Definition 3.12. A potential u on (Γ, X) is a *pair potential* if u is specified by a set of functions $\{u_k \colon X_k \rightarrow \mathbb{R}; \ u_{kj} \colon X_k \times X_j \rightarrow \mathbb{R}, k \in V, \{k, j\} \in B\}$, such that $u_{kj}(y, z) = u_{jk}(z, y)$ for all $\{k, j\} \in B$, $y \in X_k$, $z \in X_j$. ⊙

In a pair potential beside of the local weight functions u_k there are only pairwise interaction forces determined via the u_{kj}.

Definition 3.13. The transition function Q of a Markov chain with local and synchronous transition kernel according to (3.5) is a *Gibbs transition* *function with pair potential* u, *if for all* $k \in V$, $y_k \in X_k$, $x_{N(k)} \in X_{N(k)}$

$$
\begin{aligned}
\Pr\left\{\xi_k^{t+1} = y_k \mid \xi_{N(k)}^t = x_{N(k)}\right\} \\
=: Q_k\left(y_k \mid x_{N(k)}\right) \\
= \frac{\exp\left[-u_k(y_k) - \sum_{j \in N(k)} u_{kj}\left(y_k, x_j\right)\right]}{\sum_{z_k \in X_k} \exp\left[-u_k(z_k) - \sum_{j \in N(k)} u_{kj}\left(z_k, x_j\right)\right]}
\end{aligned}
\tag{3.6}
$$

holds. Here the denominator of the right-hand side gives the normalization $\sum_{y_k \in X_k} Q_k\left(y_k \mid x_{N(k)}\right) = 1$. Note that $\left(x_{N(k)}\right)_j = x_j$ holds for all $j \in N(k)$.   ⊙

We are prepared to formulate two theorems of Vasilyev and Kozlov [VK78]. These theorems give conditions for the kernel of a discrete time Markov chain (with countable state space) to be the kernel of a reversible Markov chain and to possess a stationary distribution that is Gibbsian. From the remarks after Definition 3.9 on the equivalence of Markov fields and Gibbs fields, it follows that these processes then have Markov random fields over the prescribed interaction graph as steady state, according to the theorems of Averintzev and Grimmett; see [Ave72, Gri73].

We note that the theorems of Vasilyev and Kozlov [VK78] are concerned with infinite underlying graphs. We present the version for finite graphs only where the conclusions are more stringent.

Theorem 3.14. Let $\xi = \left\{\xi^t, t = 0, 1, \ldots\right\}$, $\xi^t \colon (\Omega, \mathcal{F}, \Pr) \to (X, \mathfrak{X})$, be a discrete time Markov process with discrete state space $X := \underset{i \in V}{\times} X_i$ having local and synchronous transition probabilities with respect to the interaction graph $\Gamma = (V, B)$. Assume ξ has a strictly positive Markov transition kernel $Q = \prod_{k \in V} Q_k$, i.e., Q_k is strictly positive for all $k \in V$. Then ξ is a reversible process if and only if it is stationary and Q is a Gibbs transition function from Definition 3.13 with some pair potential u according to Definition 3.12.   ⊙

Theorem 3.15. Let $\xi = \{\xi^t, t = 0, 1, \dots\}$, $\xi^t \colon (\Omega, \mathcal{F}, \Pr) \to (X, \mathfrak{X})$, be a discrete time reversible Markov process with discrete state space $X :=$ $\underset{i \in V}{\times} X_i$ having local and synchronous transition probabilities with respect to the interaction graph $\Gamma = (V, B)$. If the transition kernel Q of ξ is a Gibbs transition function with pair potential u, then its invariant probability measure π is a Gibbs measure over (Γ, X). $\odot$

The importance of these theorems is that they provide us with structural conditions that make the Markov property in space hereditary in the sense that from the local structure with respect to Γ (which resembles the Markov property in space) of the transition kernel, the local structure of the limiting and stationary distribution follows.

This is in general not the case; see, e.g., Remark 3.34 on the situation in general queueing networks. The importance of reversibility in our class of models becomes clear now (especially since we have then (2.3)). But unfortunately enough, many of the models of practical interest are not reversible. This is in contrast to many classes dealt with in statistical physics.

3.3 LOCAL CONTROL OF INTERACTING MARKOV PROCESSES WITH GRAPH STRUCTURED STATE SPACE

In this section we describe the structure of decision models with local and synchronous internal structure. We introduce classes of admissible local strategies (policies) to control the interacting coordinates of the stochastic processes under consideration. We therefore specify the general form of the strategies from Definition 2.17 and 2.18 in a way that it is adapted to the setting of the product state space of Section 3.1 and the interacting processes of Section 3.2. For simplicity of the presentation, we assume that some uncontrolled Markov random field $\xi = (\xi^t \colon t \in \mathbb{N})$, according to Definition 3.10, is given as our starting point. ξ will be equipped in the sequel with a control structure that governs the jump transition kernel ξ to be described in Definition 3.21 below. We mainly consider controls, which are local in the sense to be specified now.

Definition 3.16 (Action spaces and local restrictions). The sequence of decision instants (control instants) is the time scale $\mathbb{N}$.

(1) The action space (set of control values) usable at control instants is $A = \underset{i \in V}{\times} A_i$ over Γ, where A_i is a set of possible actions (decisions) for vertex i. We assume that A_i is a Polish space with Borel σ-algebra $\mathfrak{A}_i$. $\mathfrak{A}$ is the Borel product-σ-algebra over A.

(2) If for the decision maker at node i at time t with history $h^t \in H^t$ the set of control actions is restricted to $A_i^t(h^t) \subseteq A_i$, we call $A_i^t(h^t)$ the set of locally admissible actions (decisions) at time t with history $h^t \in H^t$.

(3) We assume that the so defined set valued maps

$$A_i^t \colon H^t \to 2^{A_i} - \{\emptyset\}, \quad h \to A_i^t(h), \qquad i \in V,$$

depend on the local history only and are Borel measurable in the sense of Definition 2.34.

Here for a given history $h^t = \left(x^0, a^0, x^1, a^1, x^2, \ldots, x^{t-1}, a^{t-1}, x^t\right) \in H^t$ the local history $h_i^t \in H_i^t$ at node i is defined to be

$$h_i^t = \left(x_{\widetilde{N}(i)}^0, a_i^0, x_{\widetilde{N}(i)}^1, a_i^1, x_{\widetilde{N}(i)}^2, \ldots, x_{\widetilde{N}(i)}^{t-1}, a_i^{t-1}, x_{\widetilde{N}(i)}^t\right) \in H_i^t.$$

H_i^t is endowed with the trace $\mathfrak{H}_i^t$ of the product-σ-algebra on the space $\left(X_{\widetilde{N}(i)} \times A_i\right)^t \times X_{\widetilde{N}(i)}$. We also assume the sets $K_i^t = \left\{\left(h_i^t, a_i\right) \colon h_i^t \in H_i^t, a_i \in A_i^t(h^t)\right\}$ contain the graph of a measurable mapping and are Borel measurable in the trace of the product-σ-algebra $\mathfrak{K}_i^t := K_i^t \cap \left(\mathfrak{H}_i^t \times \mathfrak{A}_i\right)$.

Furthermore, the sets

$$\kappa_i^t = \left\{\left(x_{\widetilde{N}(i)}, a_i\right) \colon x_{\widetilde{N}(i)} \in X_{\widetilde{N}(i)}, a_i \in A_i^t\left(x_{\widetilde{N}(i)}\right)\right\}$$

are assumed to be Borel measurable sets of the product space $X_{\widetilde{N}(i)} \times A_i$
and $\kappa^t = \underset{i \in V}{\times} \kappa_i^t$ to be Borel measurable in the product space $\widetilde{X} \times A$, where

$$\widetilde{X} = \underset{i \in V}{\times} X_{\widetilde{N}(i)} \text{ and } \widetilde{\mathfrak{X}} = \sigma\left\{\underset{i \in V}{\times} \mathfrak{X}_{\widetilde{N}(i)}\right\}.$$

If the mappings A_i^t do not depend on t, we write $\kappa_i := \kappa_i^t$, $t \in \mathbb{N}$, and
$\kappa := \kappa^t$, $t \in \mathbb{N}$. $\odot$

Definition 3.17 (Synchronous strategies on product spaces). Let
α_i^t denote the action chosen by the decision maker at node i at time t,
$\alpha^t := \left(\alpha_i^t : i \in V\right)$ the joint decision vector at time t.

(1) A randomized strategy (policy) $\pi = \left(\pi^t : t \in \mathbb{N}\right)$ according to Definition 2.18 to control the system with interacting components and with action
space of product form is defined as vector of local policies $\pi = (\pi_i, i \in V)$,
where for node i $\pi_i = \left\{\pi_i^0, \dots, \pi_i^t, \dots\right\}$ is a sequence of transition probabilities $\pi_i^t = \pi_i^t\left(\cdot \mid x^0, a^0, \dots, x^{t-1}, a^{t-1}, x^t\right)$. So π_i^t is a probability measure
on $(A_i, \mathfrak{A}_i)$ for any $\left(x^0, a^0,, \dots, x^{t-1}, a^{t-1}, x^t\right)$ and measurably dependent
on the history $h^t = \left(x^0, a^0, \dots, x^{t-1}, a^{t-1}, x^t\right)$ of the system up to t-th
transition. We therefore have for all $B_i \in \mathfrak{A}_i$

$$\Pr\left\{\alpha_i^t \in B_i \mid \xi^0 = x^0, \alpha^0 = a^1, \dots, \xi^{t-1} = x^{t-1}, \alpha^{t-1} = a^{t-1}, \xi^t = x^t\right\}$$
$$= \pi_i^t\left(B_i \mid x^0, a^0, \dots, x^{t-1}, a^{t-1}, x^t\right). \quad (3.7)$$

(2) In parallel to the synchronous transitions and the locality of the
transition kernels, we always assume that the decision makers located at the
nodes act conditionally independent given the history of the system. This
leads to a control of the process governed by a synchronous control kernel

$$\Pr\left\{\alpha^t \in \underset{i \in V}{\times} B_i \mid \xi^0 = x^0, \alpha^0 = a^0, \dots\right.$$

$$\left. \dots, \xi^{t-1} = x^{t-1}, \alpha^{t-1} = a^{t-1}, \xi^t = x^t\right\}$$

$$= \prod_{i \in V} \Pr\left\{\alpha_i^t \in B_i \mid \xi^0 = x^0, \alpha^0 = a^0, \dots\right.$$

$$\ldots, \xi^{t-1} = x^{t-1}, \alpha^{t-1} = a^{t-1}, \xi^t = x^t\}$$
$$= \prod_{i \in V} \pi_i^t \big(B_i \mid x^0, a^0, \ldots, x^{t-1}, a^{t-1}, x^t \big),$$

$$B_i \in \mathfrak{A}_i, \quad a^s \in A, \quad x^s \in X. \quad (3.8)$$

$\odot$

Definition 3.18 (Local strategies). (1) Assume that at transition times $t = 0, 1, \ldots$, the admissible decision sets depend on the local history only according to Definition 3.16 **(3)** and that the decision α_i^t at node i is made according to the probability π_i^t on basis of the local history $h_i^t = \Big(x_{\widetilde{N}(i)}^0, a_i^0, \ldots, x_{\widetilde{N}(i)}^{t-1}, a_i^{t-1}, x_{\widetilde{N}(i)}^t \Big)$ of the states of the neighborhood $\widetilde{N}(i)$ of i and the previous decisions at i only. If $\pi_i^t \Big(A_i^t\big(h_i^t\big) \mid h_i^t \Big) = 1$, then π_i^t is said to be locally admissible, and the sequence of transition probabilities (decisions) $\pi_i = \big\{ \pi_i^t, t \in \mathbb{N} \big\}$ is called admissible local strategy for vertex i.

$\pi = (\pi_i, i \in V)$ is called admissible local strategy for the decision model.

(2) An admissible local strategy $\pi = (\pi_i, i \in V)$ is called admissible local Markov strategy if

$$\pi_i^t \Big(\cdot \mid x_{\widetilde{N}(i)}^0, a_i^0, \ldots, x_{\widetilde{N}(i)}^{t-1}, a_i^{t-1}, x_{\widetilde{N}(i)}^t \Big) = \pi_i^t \Big(\cdot \mid x_{\widetilde{N}(i)}^t \Big), \qquad i \in V.$$

Note, that whenever we deal with local Markovian strategies, we can assume that $A_i^t\big(h_i^t\big)$ depends on h_i^t only through $x_{\widetilde{N}(i)}^t$; this reduced dependence is expressed by $A_i^t\big(h_i^t\big) =: A_i^t\Big(x_{\widetilde{N}(i)}^t\Big)$.

(3) An admissible local Markov strategy $\pi = (\pi_i, i \in V)$ is called admissible local stationary (Markov) strategy if $\pi_i^{t'} \Big(\cdot \mid x_{\widetilde{N}(i)} \Big) = \pi_i^{t''} \Big(\cdot \mid x_{\widetilde{N}(i)} \Big)$, $i \in V$, for all t', t'' and all x.

(4) An admissible local stationary (Markov) strategy $\pi = (\pi_i, i \in V)$ is called admissible local stationary deterministic (nonrandomized) strategy if

$\pi_i\big(\cdot \mid x_{\widetilde{N}(i)}\big)$, $i \in V$, are one-point measures on $A_i^t\big(x_{\widetilde{N}(i)}\big)$, $i \in V$, for all $x \in X$.

The class of all admissible local strategies is denoted by LS; the subclass of admissible local Markov strategies by LS_M. By LS_S we denote the class of admissible local stationary strategies. By LS_P we denote the class of admissible local deterministic (= pure) strategies, by LS_{PM} the class of admissible local deterministic (= pure) Markov strategies (which need not be stationary), by LS_D the class of admissible local stationary (Markov) deterministic strategies.   ⊙

Remark 3.19. We have

$$LS_D \subseteq LS_S \subseteq LS_M \subseteq LS$$

and

$$LS_D \subseteq LS_{PM} \subseteq LS_P \subseteq LS.$$

Note that always $LS_{(\cdot)} \subseteq \Pi_{(\cdot)}$, see Definition 2.18 and Remark 2.19.   ⊙

Remark 3.20. We shall later consider controlled processes in continuous time and use controls that are families $\pi = \big(\pi^t \colon t \geq 0\big)$ of suitable transition kernels as randomized controls, or pure strategies which are in the Markovian case then functions $\Delta = \big(\Delta^t \colon X \to A, t \in [0, \infty)\big)$.

Without further remarks we shall use the same notation as in Remark 3.19.

The same procedure will apply if we are concerned with controlled processes in continuous time where the control and decision making is only allowed at an embedded sequence of random or deterministic time points, as, e.g., in the case of semi-Markov processes (Section 5.1) or Markov jump processes (Section 5.2).   ⊙

In our context, it is natural to assume that the control of the time dependent random field ξ results in an evolution that is governed via modified transition probabilities

$$\Pr\left\{\xi^{t+1} \in C^{t+1} \mid \xi^t = y, \left(\alpha_i^t : i \in V\right) = a\right\} = Q\left(C^{t+1} \mid y, a\right),$$

where $Q(\cdot \mid \cdot, \cdot)$ is a transition kernel from $X \times A$ to X. Note that the assumption of a local control structure is natural in our setting for defining locally interacting controlled systems. The possibility of making decisions that depend on the whole local history according to (3.7) and (3.8) however is not dispensable, because, e.g., we want the decision maker being able to learn from the observed reactions of the system responding to control actions.

Our aim is to arrive at a Markovian structure similar to Definition 3.10. The construction of the process will lead us to the assumption that the laws of motion for the system can be characterized by a time invariant set of transition probabilities. Namely, whenever the system is in state y and action a is taken, then, regardless of its history, the next state is selected according to a transition law, which depends only on (y, a).

Using this, a strategy π will define a probability measure on the space of sequences $\left(a^0, a^1, \dots\right)$ for every fixed initial state x^0. The pair (ξ, π) will be called controlled version of ξ using strategy π. The controlled process in general will not be Markovian, because the functions π_i^t, $i \in V$, depend not only on states $x_{\widetilde{N}(i)}^t$, $i \in V$, but on the previous (local) states $x_{\widetilde{N}(i)}^0, \dots, x_{\widetilde{N}(i)}^{t-1}$ as well. The Markov property is introduced in the following definition, a discussion of the principles behind will follow.

Definition 3.21 (Controlled Markov process with locally interacting synchronous components). A pair (ξ, π) is a controlled process with locally interacting synchronous components with respect to the finite interaction graph $\Gamma = (V, B)$, if $\xi = \left(\xi^t : t \in \mathbb{N}\right)$ is a process with state space $X = \underset{i \in V}{\times} X_i$ and $\pi = \left(\pi_i : i \in V\right)$ is an admissible local strategy.

The pair (ξ, π) is called controlled Markov process with locally interacting synchronous components over (Γ, X), or shortly: A (time dependent) controlled Markov random field, if the transitions of ξ are determined as follows:

For all t the $\mathrm{Pr}^{(\xi^0, \alpha^0, \ldots, \xi^{t-1}, \alpha^{t-1}, \xi^t, \alpha^t)}$-almost surely defined conditional probabilities fulfill for all

$$K \subseteq V,\ C_j \in \mathfrak{X}_j,\ j \in K,\ y \in X,\ a_j \in A_j\left(y_{\widetilde{N}(j)}\right),\ C_K = \underset{j \in K}{\times}\, C_j$$

$$\mathrm{Pr}\left\{\xi_K^{t+1} \in C_K \mid \xi^0 = x^0, \alpha^0 = a^0, \ldots \right.$$
$$\left. \ldots, \xi^{t-1} = x^{t-1}, \alpha^{t-1} = a^{t-1}, \xi^t = y, \alpha^t = a\right\}$$

$$\overset{(1)}{=} \mathrm{Pr}\left\{\xi_K^{t+1} \in C_K \mid \xi^t = y, \alpha^t = a\right\}$$

$$\overset{(2)}{=} \prod_{j \in K} \mathrm{Pr}\left\{\xi_j^{t+1} \in C_j \mid \xi^t = y, \alpha^t = a\right\}$$

$$\overset{(3)}{=} \prod_{j \in K} \mathrm{Pr}\left\{\xi_j^{t+1} \in C_j \mid \xi_{\widetilde{N}(j)}^t = y_{\widetilde{N}(j)}, \alpha_j^t = a_j\right\}$$

$$\overset{(4)}{=} \prod_{j \in K} \mathsf{Q}_j\left(C_j \mid y_{\widetilde{N}(j)}, a_j\right)$$

$$\overset{(5)}{=} \mathsf{Q}_K\left(C_K \mid y, a\right). \tag{3.9}$$

If $K = V$ we shall write $\mathsf{Q}_V(C_V \mid y, a) = \mathsf{Q}(C \mid y, a)$.

The Markov kernel $\mathsf{Q} = \prod_{j \in V} \mathsf{Q}_j$ is said to be local and synchronous.

$\odot$

Note that with the pair (ξ, π) is associated a controlled random sequence (ξ, α) as constructed in Section 2.2. This random sequence is in the present setting Markov in time. Some remarks on the modeling principle behind this definition may be in order:

- We have assumed that the first expression is well defined. Then $\overset{(1)}{=}$ is the assumption of a Markovian transition law;

- $\overset{(2)}{=}$ expresses in terms of conditional probabilities that the coordinates act synchronously;

- $\overset{(3)}{=}$ is due to the locality of the conditional probabilities;
- Assuming to have suitable transition kernels and independence of time for the one-step transition probabilities then leads to the form $\overset{(4)}{=}$ of the Markov kernel;
- $\overset{(5)}{=}$ again results from the coordinates acting synchronously and locally.

3.4 LOCAL CONTROL OF INTERACTING MARKOV PROCESSES ON GRAPHS WITH FINITE STATE AND ACTION SPACES

In this section we consider processes with state spaces and action spaces that are finite and product spaces (see [CDK05]). These have recently found many applications, e.g., in queueing network theory. A typical example is the following:

Example 3.22 (Cyclic networks). We consider a simple linear transmission line of successive transition channels numbered $1, 2, \ldots, J$ with finite capacity. Messages arrive at channel 1 and after being transmitted there they proceed to channel 2, and so on, eventually leaving the line. The transmission protocol of the network is based on a common discrete time scale.

To protect the network from overload, where the performance is degraded due to strongly increasing protocol overhead, it is assumed that a window-flow control scheme is applied: If the line carries a certain amount of traffic no further messages are admitted, and for simplicity we assume that rejected messages are lost. So by applying such a control scheme, the overload is handled outside the critical path, which the line is considered to be.

Considering the quality of service for such a transmission system, it is intuitively obvious, that, conditioned on being admitted for entrance, e.g., for such a message, the transmission time can be reduced effectively by applying an overall small admission rate. But small admission rate is equivalent to a large loss rate, creating complementary degradation of the quality of service. Therefore, to find an optimal admission policy, we should balance carefully the costs due to lost arrivals against rewards from successful transmissions within prescribed due dates.

Clearly, an optimal admission policy will incorporate the complete state description of the line at arrival instants in making the decision. But from a practical point of view such a procedure is impossible, especially in today's high speed networks. There admission decisions have to be made on the basis of the state information of the entrance node only, possibly with some information on the states of the nearest neighbor nodes.   ⊙

Steady state analysis of such a system with local admission control can be found in [PCS86]. From this, it becomes apparent that local control as investigated here often leads to models which at present are not analytically tractable; see the discussion in [DD02, Section 4]. (This is in contrast to application of some globally defined admission control policy for transmission lines [Dad97].)

Before proceeding with our general description we mention a field of research where spatiotemporal models are essentially needed as well; for more details, see [DF93, DFD98, Kno96], and the references therein. More information on Example 3.23 is presented in Section 6.1.

Example 3.23 (Diffusion of knowledge and technologies). Aspects of spatial localization that accompany competing innovations in the use of production techniques at present seem to be not well understood with respect to quantitative measures of such behavior. These aspects arise from influences described by classical location theory as well as from innovation diffusion between firms that are geographically neighbored, but produce on different levels of production techniques. Using the classical Ising models for spin systems from statistical physics [Lig85, Chapter: Stochastic Ising Models], which are random fields with pair potentials according to Definition 3.12, a simple model, where effects of technologies which are established in some locally bounded enclaves may diffuse into contiguous districts, is described in [DFD98]. These models allow us to describe to a certain extent the interaction of firms through shared knowledge and technologies. The firms are thought to be located at the nodes of a subset of $\mathbb{Z}^2$ and their evolution over time shows something like the Ising model's equilibrium behavior. This behavior is investigated in a simulation study in [DFD98, Section 6]. Although the simulations reproduce some of the aspects observed empirically, the authors point out that there is much need for more detailed models. Some problems mentioned are: Mimicking learning effects, which will lead to optimization of the firms' behavior, which is just the need for closed loop control of the system. This apparently should be modeled in a first step on a firm's local level, including the neighborhood information for that firm (*localized learning effects*, [DFD98, p. 23]). Further problems are: Feedback from the macro-state of the system, which, e.g., can be incorporated into the model by applying globally determined cost functions; allowing more than one site of the spatial lattice to change the state coordinate at the same time; and so on.

Further details on Ising models and applications in the field of knowledge diffusion are presented Example 3.41, and Section 6.1, especially in Definition 6.1.   ⊙

The problems connected with these examples are motivating for our subsequent study.

Another class of examples has a long tradition in the field of cybernetics and related areas: Automata and synchronized automata networks — we consider here synchronized networks of stochastic automata. Synchronization means that we consider systems on a common discrete time scale for the interacting automata.

Example 3.24 (Cellular Automata).

(a) We have a linearly ordered sequence of synchronized finite automata. Time scale is $\mathbb{N}$. The underlying finite undirected graph is $\Gamma = (V, B)$, with $V = \{1, 2, \ldots, S\}$ and B representing the natural neighborhoods: $N(j) = \{j - 1, j + 1\}$, $j \in \{2, \ldots, S - 1\}$, $N(1) = \{2\}$, $N(S) = \{S-1\}$. The possible states of automaton j (at position $j \in V$) are X_j, $|X_j| < \infty$, its input set (input alphabet) is I_j, $|I_j| < \infty$, and output set is O_j, $|O_j| < \infty$. The evolution of the cellular systems incorporates interactions and correlations among the automata according the graph structure prescribed by Γ. The input processes $(\alpha_j^t : t \in \mathbb{N})$ with input alphabet I_j, $j \in V$, are independent of another and independent of the history of the system.

The state process of the system of automata is

$$\xi = \left((\xi_j^t : j \in V) : t \in \mathbb{N} \right)$$

and the output process is

$$\gamma = \left((\gamma_j^t : j \in V) : t \in \mathbb{N} \right).$$

If at time t the automaton j is in state $\xi_j^t = x_j \in X_j$, and at time t an input $\alpha_j^t = y_j \in I_j$ occurs at node j, then at time $t + 1$ the internal state of the automaton j is $\xi_t^{t+1} = \tilde{x}_j$, and the output is at time $\gamma_j^{t+1} = o_j \in O_j$, with probability

$$Q_j \left(x_{j-1}, x_j, x_{j+1}, y_j; \tilde{x}_j, o_j \right)$$
$$= \mathrm{Pr} \left(\xi_j^{t+1} = \tilde{x}_j, \gamma_j^{t+1} = o_j \mid \alpha_j^t = y_j, \xi_i^t = x_i, i = j - 1, j, j + 1 \right),$$

where

$$Q_j : (X_{j-1} \times X_j \times X_{j+1} \times I_j) \times (X_j \times O_j) \longrightarrow [0, 1], \quad j \in \{2, \ldots, S-1\},$$

is a transition counting density. (Q_1, Q_S are defined similarly.)

Given the global state of the automaton at time $t - 1$ and the next inputs, we assume that the local (state/output)-values change conditionally independent over the automata. This leads to a transition structure similar to that of Definition 3.21, where we do not have the additional output function.

Prescribing an initial state $\left(\xi_j^0 \colon j \in \{1, \ldots, S\}\right)$ yields a uniquely defined stochastic behaviour of the cellular automaton.

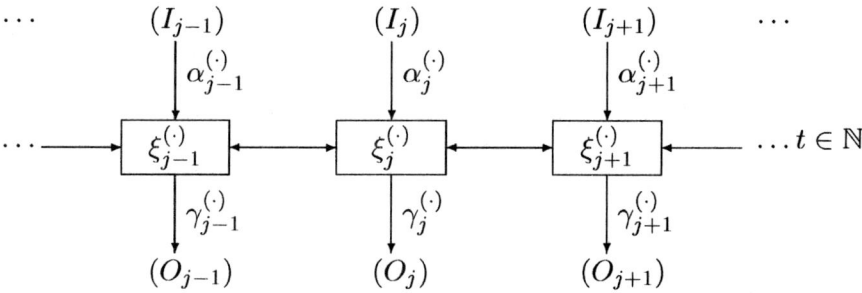

Figure 3.1: Cellular Automata

(b) In many situations a countably infinite underlying regular graph $\mathbb{Z}^d$, $d > 1$, will be adequate — or some finite subsets of $\mathbb{Z}^d$.

(c) Analogously we can model a cellular automaton, where the interaction is due to the use of common databases.

(d) If in Example (a) we have $|I_j| = 1$, $j \in \{1, \ldots, S\}$, then the input is only relevant for synchronization. This results in an autonomous cellular automaton.

(e) If we have a cellular automaton without output function, then Definition 3.21 is met precisely. □

Cellular automata as models for neural networks are an early example of models for large graph structured interacting systems. Introducing stochastic effects into the formalism of deterministic cellular automata was motivated by von Neumann [Neu56] as a feature that reflects aspects of unreliability in the local coordinates, i.e., the neurons. Such unreliability is fundamental for all organisms, and the aim of the research was to construct models that on the global level behave nearly completely reliable although the fundamental blocks of the system were unreliable.

Cellular automata nowadays provide a versatile class of models for complex interacting systems with and without control. Their range of applications covers, e.g., artificial life simulations, crystal growth, self-organization of chemical reaction-diffusion systems, vehicular traffic flow, pattern formation, and natural ecology. For a detailed overview, see [Ila02]. ⊙

Connected with our optimization problems we consider the inputs to the automata as decisions of decision makers who guide the system's behavior by independent local perturbations. The reaction of the system is dependent on the neighborhood structure and the automaton's computations result in the output and possibly an additional reward function.

3.4.1 Finite state Markov chains with locally interacting synchronous components and local control

An immediate consequence of the fundamental Definition 3.10 is for finite state space a transition mechanism of the following structure for the transition densities:

Corollary 3.25. For any time $t \in \mathbb{N}$ we have

$$\Pr\left\{\xi_K^{t+1} = x_K^{t+1} \mid \xi^t = x^t\right\} = \prod_{k \in K} \Pr\left\{\xi_k^{t+1} = x_k^{t+1} \mid \xi_{\widetilde{N}(k)}^t = x_{\widetilde{N}(k)}^t\right\},$$

for $K \subseteq V$, $x^t, x^{t+1} \in X$. ⊙

For a general discussion of such processes, see [DKT78], and especially [Vas78], and the references therein.

It is well known that in the case of finite state spaces and finite action spaces, we can restrict our search for optimal policies to the set of deterministic strategies according to Definition 2.17; see [Der70]. In Subsection 3.4.3 we study in depth strategies in this class and present in the rest of this subsection the necessary adapted definitions from Sections 3.2 and 3.3.

Assumption 3.26. (1) The sets of admissible actions are time independent; i.e., the set valued functions A^t in Definition 2.16 do not depend on t and we have $A^t(x) = A(x) \subseteq A$ for all $t \in \mathbb{N}$.

(2) The admissible set of actions for vertex $i \in V$ under system configuration (state) $x \in X$ is the set $A_i(x)$, and $A(x) = \prod_{i \in V} A_i(x)$. ⊙

The product structure of state and action spaces implies trivially that the deterministic strategies show a product structure as well.

Definition 3.27. The history dependent decisions at time t are

$$\Delta^t = \Delta^t\big(x^0, \ldots, x^t\big) = \Big\{\Delta_i^t\big(x^0, \ldots, x^t\big) : i \in V\Big\} \in \underset{i \in V}{\times} A_i, \quad t \in \mathbb{N}. \quad ⊙$$

We shall return to the more general setting in Section 3.4.5. The locality for the strategies of Definition 3.18 reads in the case of deterministic strategies as follows.

Definition 3.28. (1) For a local pure (deterministic) strategy Δ at time $t = 0, 1, \ldots$ the decision Δ_i^t for node i is made on basis of the history of the complete neighborhood $\widetilde{N}(i)$ of i only, i.e., on the basis of $x_{\widetilde{N}(i)}^0, \ldots, x_{\widetilde{N}(i)}^t$.

We therefore can consider the history dependent decision sequences $\Delta_i^t\left(x^0, \ldots, x^t\right) = \Delta_i^t\left(x_{\tilde{N}(i)}^0, \ldots, x_{\tilde{N}(i)}^t\right)$ as functions defined on the space $X_{\tilde{N}(i)}^{t+1} = \underbrace{X_{\tilde{N}(i)} \times \cdots \times X_{\tilde{N}(i)}}_{(t+1)\text{-times}}$ with values in A_i.

(2) A local pure strategy $\Delta = (\Delta_i : i \in V)$ is therefore determined by functions $\Delta_i = \{\Delta_i^t, \; t \geq 0\}$, and their values

$$\Delta_i^0\left(x_{\tilde{N}(i)}^0\right), \ldots, \Delta_i^t\left(x_{\tilde{N}(i)}^0, \ldots, x_{\tilde{N}(i)}^t\right), \ldots.$$

(3) This local pure strategy $\Delta = (\Delta_i : i \in V)$ is local Markov if for all $x^0, x^1, \ldots, x^t \in X$ $\Delta_i^t\left(x_{\tilde{N}(i)}^0, \ldots, x_{\tilde{N}(i)}^t\right) = \Delta_i^t\left(x_{\tilde{N}(i)}^t\right)$, $i \in V$, i.e., every local decision function Δ_i^t depends on the whole history only through the present local states in the neighborhood of i.

(4) The local Markov strategy $\Delta = (\Delta_i : i \in V)$ is stationary (local Markov) if $\Delta_i^{t'}\left(x_{\tilde{N}(i)}\right) \equiv \Delta_i^{t''}\left(x_{\tilde{N}(i)}\right)$, $i \in V$, for all t' and t''. Consequently, a stationary deterministic (Markov) strategy Δ is completely determined by functions Δ_i with

$$\Delta_i \colon X_{\tilde{N}(i)} \to A_i; \quad x_{\tilde{N}(i)}^t \to \Delta_i\left(x_{\tilde{N}(i)}^t\right), \qquad i \in V. \qquad \odot$$

Note that from the very definition, a deterministic stationary strategy is Markovian.

From this definition, it is natural to assume that the control of the time dependent random field ξ results in an evolution that is governed via modified transition probabilities

$$\Pr\left\{\xi^{t+1} = x^{t+1} \mid \xi_{\tilde{N}(i)}^t = x_{\tilde{N}(i)}, \left\{\Delta_i^t\left(\xi_{\tilde{N}(i)}^t\right) : i \in V\right\} = a\right\},$$

which are of product form.

We summarize the principles and modeling features to control a system with locally interacting components, which we have presented up to now in the following definition for the case of finite state and action spaces.

Definition 3.29. A pair (ξ, Δ) is called a controlled process with locally interacting synchronous components with respect to the finite interaction graph $\Gamma = (V, B)$, if

- $\xi = (\xi^t \colon t \in \mathbb{N})$ is a stochastic process with state space $X = \underset{i \in V}{\times} X_i$,
- $\Delta = (\Delta_i \colon i \in V)$ is a local pure strategy, and
- the transition kernel of ξ is determined as follows: If

$$0 < \Pr\left\{\xi^0 = x^0, \Delta^0(\xi^0) = a^0, \ldots, \xi^{t-1} = x^{t-1}, \right.$$
$$\left. \Delta^{t-1}(\xi^0, \ldots, \xi^{t-1}) = a^{t-1}, \xi^t = y, \Delta^t(\xi^0, \ldots, \xi^t) = a\right\},$$

then

$$\Pr\left\{\xi_K^{t+1} = x_K \mid \xi^0 = x^0, \Delta^0(\xi^0) = a^0, \ldots\right.$$
$$\left. \ldots, \xi^{t-1} = x^{t-1}, \Delta^{t-1}(\xi^0, \ldots, \xi^{t-1}) = a^{t-1}, \xi^t = y, \Delta^t(\xi^0, \ldots, \xi^t) = a\right\}$$
$$\overset{(1)}{=} \Pr\left\{\xi_K^{t+1} = x_K \mid \xi^0 = x^0, \ldots, \xi^{t-1} = x^{t-1}, \xi^t = y, \Delta^t(\xi^0, \ldots, \xi^t) = a\right\}$$
$$= \Pr\left\{\xi_K^{t+1} = x_K \mid \xi^t = y, \Delta^t(\xi^0, \ldots, \xi^t) = a\right\}$$
$$= \prod_{j \in K} \Pr\left\{\xi_j^{t+1} = x_j \mid \xi^t = y, \Delta^t(\xi^0, \ldots, \xi^t) = a\right\}$$
$$= \prod_{j \in K} \Pr\left\{\xi_j^{t+1} = x_j \mid \xi_{\widetilde{N}(j)}^t = y_{\widetilde{N}(j)}, \Delta_j^t\left(\xi_{\widetilde{N}(j)}^0, \ldots, \xi_{\widetilde{N}(j)}^t\right) = a_j\right\}$$
$$= \prod_{j \in K} Q_j\left(x_j \mid y_{\widetilde{N}(j)}, a_j\right) = Q_K(x_K \mid y, a), \quad K \subseteq V, \ y \in X, \ a \in A(y),$$

where the $Q_j\left(x_j \mid y_{\widetilde{N}(j)}, a_j\right)$ are locally defined transition kernels that determine the time invariant laws of motion for the system. We shall write for $K = V$: $Q(x \mid y, a) = \Pr\left\{\xi^{t+1} = x \mid \xi^t = y, \Delta^t = a\right\}$.   ⊙

Here $\overset{(1)}{=}$ is due to the deterministic strategy; for a comment on the further structure for the transition mechanism we refer to (3.9) and the remarks thereafter. Obviously, $\sum_{x \in X} Q(x \mid y, a) = 1$, $y \in X$, $a \in A$.

Remark 3.30. If Δ in the setting of Definition 3.29 is a stationary (Markov) strategy, $\Delta^t = \Delta^{t'}$ for all $t, t' \in \mathbb{N}$, with time invariant admissible decision sets and $A^t(x) = A(x) = \underset{i \in V}{\times} A_i\left(x_{\widetilde{N}(i)}\right)$ for all t, then ξ is a homogeneous Markov chain, and (because Δ is local,) according to Definition 3.10, a time dependent Markov random field. We call such pair (ξ, Δ) (or the process ξ) a controlled Markov process with locally interacting synchronous components with respect to the finite interaction graph $\Gamma = (V, B)$, or shortly, a time dependent controlled Markov random field.$\odot$

It remains to specify the cost structure for the system. We consider here cost functions that are stationary in time according to (2.5), and only deterministic strategies Δ. Therefore, if at time $t \in \mathbb{N}$ the system is in state $\xi^t = x^t$ and a decision for action $\alpha^t = a^t$ is made a (one-step) cost $r(x^t, a^t) \geq 0$ is incurred to the system. We consider the (minimax) cost criterion (2.6). Due to the deterministic decision rules Δ the average expected cost up to time T when ξ is started with $\xi^0 = x^0$ and strategy Δ is applied is

$$
\begin{aligned}
\rho(x^0, \Delta) &= \limsup_{T \to \infty} \mathsf{E}^{\Delta}_{x^0} \frac{1}{T+1} \sum_{t=0}^{T} r(\xi^t, \Delta^t) \\
&:= \limsup_{T \to \infty} \mathsf{E}^{\Delta}_{x^0} \frac{1}{T+1} \sum_{t=0}^{T} r\left(\xi^t, \Delta^t(\xi^0, \ldots, \xi^t)\right),
\end{aligned}
\tag{3.10}
$$

where $\mathsf{E}^{\Delta}_{x^0}$ is expectation associated with the controlled process (ξ, Δ) if $\xi^0 = x^0$.

Note that we suppressed the dependency of Δ^t on $(\xi^0, \ldots, \xi^t)$, which we shall do similarly in the following as well. The problem is to find a strategy Δ that minimizes the asymptotic average expected costs (3.10). We consider the problem of finding optimal strategies with respect to the (minimax) cost criterion from Definition 2.21 within LS_P only.

Definition 3.31. A strategy $\Delta^\star \in LS_P$ is called optimal within the class LS_P of admissible deterministic (= pure) local strategies if

$$\rho(x^0, \Delta^\star) = \inf_{\Delta \in LS_P} \rho(x^0, \Delta)$$

for all $x^0 \in X$. $\odot$

3.4.2   Cyclic networks as random fields

We describe in this subsection an important example from queueing network theory that has a transition mechanism which is obviously local and synchronous. However, it turns out that there arise difficulties with fitting the model into the definitions of the previous subsection. We will demonstrate how to overcome these difficulties.

Example 3.32 (Cyclic network). (This is a continuation of Example 3.22.) Closed linear networks (cyclic networks) arise, e.g., if window-flow control is applied for congestion control in circuit switching networks. Such flow control protocols are applied as well in high speed transmission networks. Under the ATM protocol, these networks show an internal generic discrete time scale.

As a model for these networks we consider a closed cycle of single server nodes numbered $1, 2, \ldots, J$, $J \geq 2$. Each node has ample waiting room, and the service discipline is first-come-first-served (FCFS). K undistinguishable customers are cycling in the system, $K \geq 1$. A customer leaving node j immediately proceeds to node $j + 1$, $j = 1, \ldots, J$; we set $J + 1 := 1$, $1 - 1 := J$.

The evolution over time of the system is described by a discrete time stochastic process $\xi = \left(\xi^t : t \in \{0, 1, \dots\}\right)$ with state space

$$S(K, J) = \left\{(x_1, \dots, x_J) \in \mathbb{N}^J : x_1 + \dots + x_J = K\right\}.$$

ξ records the development of the joint queue length vector, i.e., $\xi^t = \left(\xi_1^t, \dots, \xi_J^t\right) = (x_1, \dots, x_J)$ indicates that at time t there are x_j customers present at node j, including the customer just in service, if any, $j = 1, \dots, J$. The following assumptions imply that ξ is an ergodic Markov chain.

The nodes operate as independent Bernoulli servers under FCFS: if at time $t \in \mathbb{N}$ at node j a customer is in service and if there are $h - 1 \geq 0$ other customers waiting, then in this time segment her service ends with probability $p_j(h) \in (0, 1)$. She then leaves node j at time $(t + 1)-$ and arrives at node $j + 1$ just before time $(t + 1)$. With probability $q_j(h) := 1 - p_j(h)$ this customer will stay at node j, requesting for at least one further quantum of service time there. Given the actual local queue length at some node, the decision of a customer whether to stay there or to leave is made independently of anything else. A customer leaving node j at time $(t + 1)-$ will at time $t + 1$ either join the queue of node $j + 1$ or enter service immediately — if node $j + 1$ was idle during $[t, t + 1)$ or at time $(t + 1)-$ the only customer present at node $j + 1$ has left this node.

If at some node at the *same time point* a departure and an arrival occur, we always assume that the departure event takes place first and both events happen just before the clock counts. (For information on the *Departure before Arrival-rule* (D/A), see [GH92], and for information on the *late arrival rule*, see [Hun83]).

We consider for the system the interaction graph $\Gamma = (V, B)$ with $V = \{1, 2, \dots, J\}$, $B = \left\{\{j, j + 1\} : j = 1, \dots, J\right\}$ (where $J + 1 := 1$), which implies that for any node j, we have $N(j) = \{j - 1, j + 1\}$. With respect to this graph, ξ fulfills part of the properties of a time dependent Markov random field according to Definition 3.10. The transition probabilities of ξ obviously fulfill the locality condition (3.3) but they are not synchronous according to (3.4). $\odot$

We have the following theorem (see [Dad97, Theorem 1]).

Theorem 3.33 (Steady state). The joint queue length process ξ is ergodic on $S(K,J)$ and its unique limiting and stationary distribution is $\pi^{K,J} = \left(\pi^{K,J}(x)\colon x \in S(K,J)\right)$ given by

$$\pi^{K,J}(x_1,\ldots,x_J) = \prod_{j=1}^{J} \left(\frac{\prod_{h=1}^{x_j-1} q_j(h)}{\prod_{h=1}^{x_j} p_j(h)} \right) G(K,J)^{-1},$$

$$(x_1,\ldots,x_J) \in S(K,J), \quad (3.11)$$

where $G(K,J)$ is the norming constant. ⊙

Remark 3.34. The steady state distribution (3.11) is said to be of product form. Note that $\pi^{K,J}$ is not a Markov random field with respect to Γ. This observation points out that in general, the one-dimensional marginals in time of a time dependent Markov random field according to Definition 3.10 are not Markov random fields according to Definition 3.2. Note that this holds in Example 3.32, although $\pi^{K,J}$ is a conditional distribution of a vector with independent coordinates. ⊙

Although the network process of Example 3.32 acts in an obvious sense locally with respect to the natural neighborhood structure, the transition probabilities of the embedded jump chain are not locally determined and synchronized in the sense of Definition 3.10. This is a result of the fact that the evolution of the network is determined by jumps of customers in which two nodes are always concerned with in a strongly dependent way.

So in contrast to Example 3.24 where the standard description of the cellular automaton makes it obvious that (3.3) and (3.4) hold for the transition kernel, the standard Markovian description for the queueing networks yields transition kernels that do *not* show these properties.

This is seemingly contrary to our intuition, but a simple matter of the underlying state description.

On the other hand, the network obviously acts synchronized in a general sense and locally with respect to the given graph. We therefore describe in Example 3.35 a supplementary variable technique using an extended state space, which makes the time development of the system in a formal sense fulfilling the properties of a Markov jump random field.

The main idea is to include for each node (service station) j into the local state description the local queue length x_j there, *and* a decision value s_j whether at the *next* jump instant of the system the ongoing service expires $(= 1)$ or is continued for at least one further period $(= 0)$.

Example 3.35 (Supplemented cyclic network). We continue the discussion of the cyclic network of Example 3.32. The main difference to the description on page 75 is that we introduce a new state space and new state variables.

The evolution over time of the system is described by a discrete time stochastic process

$$\xi = \left(\xi^t = \left(\nu^t, \vartheta^t \right) = \left(\nu_1^t, \vartheta_1^t; \ldots; \nu_J^t, \vartheta_J^t \right) : t \in \{0, 1, \ldots\} \right)$$

with state space

$$ST(K, J) = \left\{ (x_1, s_1; \ldots; x_J, s_J) \in \left(\mathbb{N} \times \{0, 1\} \right)^J : x_1 + \ldots + x_J = K \right\}.$$

The ν_j^t values record the development of the queue length at node j, while the supplementary variable ϑ_j^t indicates whether at the *next* time instant of the system the ongoing service, if any, expires $(\vartheta_j^t = 1)$ or is continued for at least one further period $(\vartheta_j^t = 0)$.

The time evolution of the network is as follows: If at time $t \in \mathbb{N}$ the supplemented system state is

$$\xi^t = \left(\left(\nu_j^t, \vartheta_j^t \right) : j = 1, \ldots, J \right)$$
$$= \left(x_1, s_1; \ldots; x_{j-1}, s_{j-1}; x_j, s_j; \ldots; x_J, s_J \right),$$

then the system stays in this state for one time unit. When this time unit expires, then at each node j in a first step the queue length is updated according to the values

$$\left\{ \left(\nu_j^t, \vartheta_i^t \right) = (x_i, s_i) \colon i \in \tilde{N}(j) \right\}. \tag{3.12}$$

This yields (with $1 - 1 =: J$)

$$\nu_j^{t+1} = \left(\nu_j^t - \vartheta_j^t \right)^+ + \vartheta_{j-1}^t \cdot \mathbb{1}\left(\nu_{j-1}^t > 0 \right)$$
$$= (x_j - s_j)^+ + s_{j-1} \cdot \mathbb{1}\left(x_{j-1} > 0 \right)$$

with probability one.

Given the local states of the neighborhood of j according to (3.12), it is then decided independent of other details of the network's history whether an ongoing service (if any, i.e., if $\xi^{t+1} > 0$) at the next but one time step will be (successfully) finished. We have with probability

$$\mathrm{Pr}\left(\vartheta^{t+1} = 1 \mid \left(\nu_j^t, \vartheta_i^t \right) = (x_i, s_i) \colon i \in \tilde{N}(j) \right)$$
$$= p_j\left((x_j - t_j)^+ + t_{j-1} \cdot \mathbb{1}\left(x_{j-1} > 0 \right) \right),$$

and

$$\mathrm{Pr}\left(\vartheta^{t+1} = 0 \mid \left(\nu_j^t, \vartheta_i^t \right) = (x_i, s_i) \colon i \in \tilde{N}(j) \right)$$
$$= q_j\left((x_j - t_j)^+ + t_{j-1} \cdot \mathbb{1}\left(x_{j-1} > 0 \right) \right)$$
$$= 1 - p_j\left((x_j - t_j)^+ + t_{j-1} \cdot \mathbb{1}\left(x_{j-1} > 0 \right) \right),$$

and the respective decisions are conditionally independent given the states in the neighborhood. ⊙

Corollary 3.36. The Markov chain of the supplemented cyclic network process ξ from Example 3.35 fulfills the locality property (3.3) and the synchronization property (3.4) from Definition 3.10, and so it is a Markov jump random field as defined there. ⊙

Proof: Updating of the queue length according to the values $\left\{ \left(\nu_j^t, \vartheta_i^t\right) = (x_i, s_i) \colon i \in \widetilde{N}(j) \right\}$ is purely deterministic at each node, interfering only with neighboring nodes. This is due to the selected state space, and from the structure of the underlying graph conditioned ξ^t, this updating is independent. So given ξ^t we have conditional independence of the ν_j^{t+1} and obviously the required locality. The transition decisions for the ϑ_j^t are independent of the network's history given $\left\{ \left(\nu_j^t, \vartheta_i^t\right) = (x_i, s_i) \colon i \in \widetilde{N}(j) \right\}$ and conditionally on this independent of another. This yields the required conditional independence of the synchronization property. The other properties are obvious. $\qquad\qquad\qquad\qquad\qquad\qquad\qquad\qquad\qquad$ $\square$

There is an appealing further interpretation connected with the supplemented cyclic network process from Example 3.35, which is of special interest with respect to our optimization project for time dependent Markov fields.

Example 3.37 (Controlled supplemented cyclic network). We continue the discussion of Example 3.35, recall the state space $ST(K, J)$ defined there and the supplemented Markov chain

$$\xi = \left(\xi^t = \left(\nu^t, \vartheta^t\right) = \left(\nu_1^t, \vartheta_1^t; \ldots; \nu_J^t, \vartheta_J^t\right) \colon t \in \{0, 1, \ldots\} \right).$$

We now introduce at every station (node) a controller who decides whether an ongoing service (if any, i.e., if $\xi^{t+1} > 0$) at the next but one time step will be (successfully) finished or not. The respective decision variables α_j^t are drawn according to some admissible local stationary Markov strategy π (see Definition 3.18 **(3)**) with the probabilities

$$\pi_j \left(1 \mid (x_i, s_i) \colon i \in \widetilde{N}(j) \right)$$
$$= \Pr \left(\alpha_j^t = 1 \mid \left(\nu_j^t, \vartheta_i^t\right) = (x_i, s_i) \colon i \in \widetilde{N}(j) \right)$$
$$= p_j \left(\left(x_j - t_j\right)^+ + t_{j-1} \cdot \mathbb{1}\left(x_{j-1} > 0\right) \right),$$

and

$$\pi_j\left(0 \mid (x_i, s_i)\colon i \in \widetilde{N}(j)\right)$$
$$= \Pr\left(\alpha_j^t = 0 \mid \left(\nu_j^t, \vartheta_i^t\right) = (x_i, s_i)\colon i \in \widetilde{N}(j)\right)$$
$$= q_j\left(\left(x_j - t_j\right)^+ + t_{j-1} \cdot \mathbb{1}\left(x_{j-1} > 0\right)\right)$$
$$= 1 - p_j\left(\left(x_j - t_j\right)^+ + t_{j-1} \cdot \mathbb{1}\left(x_{j-1} > 0\right)\right),$$

and the respective decisions are conditionally independent given the states in the neighborhood.

Then the so defined controlled Markov process (ξ, π) has one step transition kernel of the form (3.9) from Definition 3.21 because given $\left(\xi^t, \alpha^t\right)$, the value ξ^{t+1} is completely defined (deterministic). The locality is obvious. $\odot$

We summarize the description in Example 3.37.

Corollary 3.38. The controlled process (ξ, π) with admissible local stationary Markov strategy π and controlled supplemented cyclic network process ξ from Example 3.37 is a controlled Markov process with locally interacting synchronous components according to Definition 3.21; i.e., the controlled one-step transition kernel fulfills (3.9). $\odot$

3.4.3 Optimality of deterministic Markovian strategies in the class of local strategies

Similar to the case of Markov chains with finite state space and controls (see [VS64] or [Der70]), it is possible to prove that there exists an optimal strategy in the class of stationary local Markov strategies, which are deterministic.

Theorem 3.39. Consider a controlled processes (ξ, Δ) with locally interacting synchronous components with respect to an interaction graph $\Gamma = (V, B)$ according to Definition 3.10 with finite state space X of ξ, and finite control space A. Let the sets of admissible actions $A^t(\cdot)$ be independent of t. Then in the class LS_P of admissible deterministic local strategies, there exists an optimal strategy in the sense of Definition 3.31, that is in the class LS_D of stationary Markovian deterministic policies. $\odot$

Proof: Our proof follows the lines of the proof of the similar theorem for standard controlled Markov chains in [VS64] or [Der70, Chapter 3].

We first evaluate the process and the associated costs under the total expected discounted costs criterion **(I)**. By compactness arguments we ensure the existence of optimal strategies **(a)**. From construction of sequences of suitable near optimal policies we obtain via a diagonalization argument an optimal Markovian policy **(b)**. Invoking the optimality equation we show that in the class of Markovian policies even a stationary optimal policy exists **(c)**. Applying Abelian theorems we finally transform this result to the case of long time average expected cost criterion **(II)**.

(I) Discounted costs: For every strategy $\Delta \in LS_P$ we define for discount factor $\beta \in (0, 1)$ the total expected discounted costs under Δ when starting in y by

$$\mathfrak{R}_y^\Delta(\beta) = \mathsf{E}_y^\Delta \sum_{t=0}^{\infty} \beta^t r\big(\xi^t, \Delta^t\big),$$

and by $\mathfrak{R}_y(\beta) = \inf_{\Delta \in LS_P} \mathfrak{R}_y^\Delta(\beta)$ the infimum over all admissible local pure policies.

(a) Existence: Because X and A are finite, similar to the procedure in [VS64] (see [Der70, Chapter 3, Corollary 1 and the preceding lemmas], as well), it may be shown that for any $\beta \in (0, 1)$ the functional $\mathfrak{R}_y^\Delta(\beta)$ attains its minimal value on LS_P, i.e., there exists

$$\Delta^\star = \Delta^\star(\beta) = \left\{ \overset{\star}{\Delta}{}^0(\beta), \overset{\star}{\Delta}{}^1(\beta) \ldots, \overset{\star}{\Delta}{}^t(\beta), \ldots \right\} \in LS_P:$$

$$\Delta^\star(\beta) = \left\{ \Delta_i^\star(\beta), \ i \in V \right\},$$
$$\Delta_i^\star(\beta) = \left\{ \overset{\star}{\Delta}{}_i^0(\beta), \ldots, \overset{\star}{\Delta}{}_i^t(\beta), \ldots \right\},$$

where $\overset{\star}{\Delta}{}_i^t(\beta)\left(x_{\widetilde{N}(i)}^0, \ldots, x_{\widetilde{N}(i)}^t \right) \in A^t\left(x_{\widetilde{N}(i)}^t \right)$, such that for all $y \in X$

$$\mathfrak{R}_y^{\Delta^\star}(\beta) = \mathfrak{R}_y(\beta).$$

The proof utilizes the compactness of the space of admissible policies (from finiteness of the action spaces and Tikhonov's theorem) and the continuity of the cost functional (under the discrete topology).

(Note that we wrote $\mathfrak{R}_y^{\Delta^\star}(\beta)$ for $\mathfrak{R}_y^{\Delta^\star(\beta)}(\beta)$. If from the context there is no ambiguity, we shall treat similar expressions in the same way.)

Now we show that for fixed β and every strategy $\Delta^\star(\beta)$, which is optimal under the total expected discounted cost criterion $\mathfrak{R}_y^\Delta(\beta)$, there exists some stationary Markov strategy $\widetilde{\Delta} = \widetilde{\Delta}(\beta)$, which yields the same total discounted expected cost as $\Delta^\star(\beta)$, when applied to control the system:

$$\mathfrak{R}_y^{\widetilde{\Delta}}(\beta) = \mathfrak{R}_y^{\Delta^\star}(\beta).$$

(We consider $\widetilde{\Delta}(\beta)$ therefore to be equivalent to $\Delta^\star(\beta)$ in the sense of generating the same total discounted expected cost.)

(b) Markov property: The strategy $\Delta^\star(\beta)$ is characterized by the following property:

$$\mathfrak{R}_y^{\Delta^\star}(\beta) \leq \mathfrak{R}_y^\Delta(\beta), \qquad \Delta \in LS_P, \quad \forall \, y \in X. \tag{3.13}$$

Using conditional expectations we obtain for any $p \geq 0$

$$\mathfrak{R}_{x^0}^{\Delta}(\beta) = \mathsf{E}_{x^0}^{\Delta} \left\{ \sum_{t=0}^{p} \beta^t r\left(\xi^t, \Delta^t\right) + \right.$$

$$\left. + \mathsf{E}^{\Delta} \left[\sum_{t=p+1}^{\infty} \beta^t r\left(\xi^t, \Delta^t\right) \mid \xi^0 = x^0, \ldots, \xi^p = x^p \right] \right\}.$$

We apply Bellman's principle and will show: If $x^0, \ldots, x^p$ are the first $p+1$ observations of the process $\left(\xi, \Delta^\star(\beta)\right)$, then for any strategy $\Delta(\beta) \in LS_P$

$$\mathsf{E}^{\Delta^\star} \left\{ \sum_{t=p+1}^{\infty} \beta^t r\left(\xi^t, \overset{\star}{\Delta}{}^t\right) \mid \xi^0 = x^0, \ldots, \xi^p = x^p \right\}$$

$$\leq \mathsf{E}^{\Delta} \left\{ \sum_{t=p+1}^{\infty} \beta^t r\left(\xi^t, \Delta^t\right) \mid \xi^0 = x^0, \ldots, \xi^p = x^p \right\} \quad (3.14)$$

holds almost surely with respect to the process measure of $\left(\xi, \Delta^\star(\beta)\right)$.

Indeed, let $\mathfrak{X}_p$ denote the sigma-field generated by $\left(\xi^0, \ldots, \xi^p\right)$. Suppose, there exists a strategy $\bar{\Delta} \in LS_P$ and some set $\mathfrak{M}_p = \{\xi^0 = x^0, \ldots, \xi^p = x^p\} \in \mathfrak{X}_p$, with positive $\mathrm{Pr}_{x^0}^{\Delta^\star}$-measure of the process $\left(\xi, \Delta^\star(\beta)\right)$, such that

$$\mathsf{E}^{\Delta^\star} \left\{ \sum_{t=p+1}^{\infty} \beta^t r\left(\xi^t, \overset{\star}{\Delta}{}^t\right) \mid \xi^0 = x^0, \ldots, \xi^p = x^p \right\}$$

$$> \mathsf{E}^{\bar{\Delta}} \left\{ \sum_{t=p+1}^{\infty} \beta^t r\left(\xi^t, \bar{\Delta}^t\right) \mid \xi^0 = x^0, \ldots, \xi^p = x^p \right\},$$

and the inequality (3.14) holds on the complementary set $\overline{\mathfrak{M}}_p$ of $\mathfrak{M}_p$. We then can construct a new strategy $\widetilde{\Delta}(\beta) = \{\widetilde{\Delta}_i(\beta), \ i \in V\}$,

$\widetilde{\Delta}_i(\beta) = \{\widetilde{\Delta}_i^t(\beta), \ t \geq 0\}$, by (for simplicity of presentation we omit the dependence on β)

$$\widetilde{\Delta}_i^j\left(x_{\widetilde{N}(i)}^0, \ldots, x_{\widetilde{N}(i)}^j\right) = \overset{\star}{\Delta}_i^j\left(x_{\widetilde{N}(i)}^0, \ldots, x_{\widetilde{N}(i)}^j\right), \qquad j \leq p,$$

and for $j \geq 1$

$$\widetilde{\Delta}_i^{p+j}\left(x_{\widetilde{N}(i)}^0, \ldots, x_{\widetilde{N}(i)}^p, \ldots, x_{\widetilde{N}(i)}^{p+j}\right)$$

$$= \begin{cases} \bar{\Delta}_i^{p+j}\left(x_{\widetilde{N}(i)}^0, \ldots, x_{\widetilde{N}(i)}^p, \ldots, x_{\widetilde{N}(i)}^{p+j}\right), & (x^0, \ldots, x^p) \in \mathfrak{M}_p, \\ \overset{\star}{\Delta}_i^{p+j}\left(x_{\widetilde{N}(i)}^0, \ldots, x_{\widetilde{N}(i)}^p, \ldots, x_{\widetilde{N}(i)}^{p+j}\right), & (x^0, \ldots, x^p) \in \overline{\mathfrak{M}}_p. \end{cases}$$

Thus, for the strategy $\widetilde{\Delta}(\beta)$, which is from its very construction in LS_P, we have

$$\mathfrak{R}_{x^0}^{\widetilde{\Delta}}(\beta) = \mathsf{E}_{x^0}^{\widetilde{\Delta}}\left\{ \sum_{t=0}^{p} \beta^t r\left(\xi^t, \widetilde{\Delta}^t\right) + \right.$$

$$\left. + \mathsf{E}^{\widetilde{\Delta}}\left[\sum_{t=p+1}^{\infty} \beta^t r\left(\xi^t, \widetilde{\Delta}^t\right) \mid \xi^0 = x^0, \ldots, \xi^p = x^p \right] \right\}$$

$$= \mathsf{E}_{x^0}^{\widetilde{\Delta}} \sum_{t=0}^{p} \beta^t r\left(\xi^t, \widetilde{\Delta}^t\right) +$$

$$+ \int_{\mathfrak{M}_p} \mathsf{E}^{\widetilde{\Delta}}\left[\sum_{t=p+1}^{\infty} \beta^t r\left(\xi^t, \widetilde{\Delta}^t\right) \mid \xi^0 = x^0, \ldots, \xi^p = x^p \right] \times$$

$$\times d\Pr_{x^0}^{\Delta^\star}\left\{\xi^0 = x^0, \ldots, \xi^p = x^p\right\} +$$

$$+ \int_{\overline{\mathfrak{M}}_p} \mathsf{E}^{\widetilde{\Delta}}\left[\sum_{t=p+1}^{\infty} \beta^t r\left(\xi^t, \widetilde{\Delta}^t\right) \mid \xi^0 = x^0, \ldots, \xi^p = x^p \right] \times$$

$$\times d\Pr_{x^0}^{\Delta^\star}\left\{\xi^0 = x^0, \ldots, \xi^p = x^p\right\}$$

$$< \mathsf{E}_{x^0}^{\Delta^\star} \sum_{t=0}^{p} \beta^t r\left(\xi^t, \overset{\star}{\Delta}^t\right) +$$

$$+ \int_{\mathfrak{M}_p} \mathsf{E}^{\Delta^\star} \left[\sum_{t=p+1}^{\infty} \beta^t r\left(\xi^t, \overset{\star}{\Delta}{}^t\right) \mid \xi^0 = x^0, \dots, \xi^p = x^p \right] \times$$

$$\times d\Pr_{x^0}^{\Delta^\star} \left\{ \xi^0 = x^0, \dots, \xi^p = x^p \right\} +$$

$$+ \int_{\widetilde{\mathfrak{M}}_p} \mathsf{E}^{\Delta^\star} \left[\sum_{t=p+1}^{\infty} \beta^t r\left(\xi^t, \overset{\star}{\Delta}{}^t\right) \mid \xi^0 = x^0, \dots, \xi^p = x^p \right] \times$$

$$\times d\Pr_{x^0}^{\Delta^\star} \left\{ \xi^0 = x^0, \dots, \xi^p = x^p \right\}$$

$$= \mathfrak{R}_{x^0}^{\Delta^\star}(\beta),$$

i.e., $\mathfrak{R}_{x^0}^{\widetilde{\Delta}}(\beta) < \mathfrak{R}_{x^0}^{\Delta^\star}(\beta)$, which contradicts inequality (3.13). Therefore we henceforth assume that (3.14) is fulfilled for the optimal strategy $\Delta^\star(\beta)$ for any realization $(x^0, \dots, x^p)$ with nonzero probability under $\Pr_{x^0}^{\Delta^\star}$ (because X is finite such realizations exist). Setting

$$\Phi^{\Delta^\star}(x^0, \dots, x^p)$$

$$= \mathsf{E}^{\Delta^\star} \left\{ \sum_{t=p+1}^{\infty} \beta^t r\left(\xi^t, \overset{\star}{\Delta}{}^t\right) \mid \xi^0 = x^0, \dots, \xi^p = x^p \right\}, \quad (3.15)$$

similar to [VS64, Lemma 2], we shall prove that the function $\Phi^{\Delta^\star}(x^0, \dots, x^p)$ depends on the whole history only through x^p; i.e., for all $x^m, y^m, m = 0, 1, \dots, p-1$, we have $\Phi^{\Delta^\star}(x^0, \dots, x^{p-1}, x^p) = \Phi^{\Delta^\star}(y^0, \dots, y^{p-1}, x^p)$.

We fix time m and consider all finite time realizations $(x^0, \dots, x^{m-1}, x^m)$ of ξ until time m, which at time m attain the prescribed value x^m, and $\Phi^{\Delta^\star}(x^0, \dots, x^{m-1}, x^m)$ according to (3.15). Because X is finite, we find only a finite number of such realizations and therefore $\Phi^{\Delta^\star}(x^0, \dots, x^{m-1}, x^m)$ attains in, say, $(\hat{x}^0, \dots, \hat{x}^{m-1}, x^m) := (\hat{x}^0(x^m), \dots, \hat{x}^{m-1}(x^m), x^m)$, its minimum on the set of all such realizations, and this sequence occurs with positive $\Pr_{\hat{x}^0}^{\Delta^\star}$-measure.

So for any other realization $(y^0, \dots, y^{m-1}, x^m)$ that occurs with positive $\Pr_{y^0}^{\Delta^\star}$-measure, we have

$$\Phi^{\Delta^\star}(\hat{x}^0, \dots, \hat{x}^{m-1}, x^m) \le \Phi^{\Delta^\star}(y^0, \dots, y^{m-1}, x^m). \quad (3.16)$$

We show that in (3.16), equality holds. Assume we have strict inequality

$$\Phi^{\Delta^\star}\left(\hat{x}^0,\ldots,\hat{x}^{m-1},x^m\right) < \Phi^{\Delta^\star}\left(y^0,\ldots,y^{m-1},x^m\right)$$

for some realization $\left(y^0,\ldots,y^{m-1},x^m\right)$ of the process. Then we construct a new admissible strategy $\Delta^{\star\star}$, which acts as $\Delta^\star$ until time m, and which after time m acts on any sequence $\left(x^0,\ldots,x^{m-1},x^m,x^{m+1},\ldots,x^{m+k}\right)$ of states in the same way as $\Delta^\star$ would do when $\left(\hat{x}^0,\ldots,\hat{x}^{m-1},x^m\right)$ would have happened up to time m:

$$\overset{\star\star}{\Delta}{}^{m+k}\left(x^0,\ldots,x^{m-1},x^m,x^{m+1},\ldots,x^{m+k}\right)$$
$$:= \overset{\star}{\Delta}{}^{m+k}\left(\hat{x}^0,\ldots,\hat{x}^{m-1},x^m,x^{m+1},\ldots,x^{m+k}\right).$$

For the strategy $\Delta^{\star\star}$ we therefore obtain

$$E^{\Delta^{\star\star}}\left\{\sum_{t=p+1}^{\infty}\beta^t r\left(\xi^t,\overset{\star\star}{\Delta}{}^t\right) \mid \xi^0 = x^0,\ldots,\xi^p = x^p\right\}$$
$$< E^{\Delta^\star}\left\{\sum_{t=p+1}^{\infty}\beta^t r\left(\xi^t,\overset{\star}{\Delta}{}^t\right) \mid \xi^0 = x^0,\ldots,\xi^p = x^p\right\},$$

which contradicts (3.14), and thus we have proved that $\Phi^{\Delta^\star}\left(x^0,\ldots,x^m\right)$ depends on the whole history only through x^m.

For any realization $\left(x^0,\ldots,x^{m-1},x^m,x^{m+1},\ldots,x^{m+k},\ldots\right)$ of ξ (with our fixed x^m at time m) we therefore apply from time m on the control decisions

$$\overset{\star}{\Delta}{}^{m+k}\left(\hat{x}^0,\ldots,\hat{x}^{m-1},x^m,x^{m+1},\ldots,x^{m+k}\right) \text{ instead of}$$
$$\overset{\star}{\Delta}{}^{m+k}\left(x^0,\ldots,x^{m-1},x^m,x^{m+1},\ldots,x^{m+k}\right), \qquad k \geq 1.$$

As we have shown, this does not change the value $\Phi^{\Delta^\star}\left(x^0,\ldots,x^m\right)$. Therefore the strategy $\Delta^{(m)}$, defined to be equal to $\Delta^\star$ until time m, and then defined by

$$\overset{(m)}{\Delta}{}^{m+k}\left(x^0,\ldots,x^{m-1},x^m,x^{m+1},\ldots,x^{m+k}\right)$$

$$:= \overset{(\star)}{\Delta}{}^{m+k}\left(\hat{x}^0,\ldots,\hat{x}^{m-1},x^m,x^{m+1},\ldots,x^{m+k}\right), \qquad k\geq 0,$$

yields the same conditional expectation as the optimal control $\Delta^\star$:

$$\Phi^{\Delta^{(m)}}\left(x^0,\ldots,x^m\right) = \Phi^{\Delta^\star}\left(x^0,\ldots,x^m\right).$$

Note, that by definition $\Delta^{(m)}$ is a local strategy because of the locality of $\Delta^\star$ and that

$$\overset{(m)}{\Delta}{}^{m+k}\left(x^0,\ldots,x^{m-1},x^m,x^{m+1},\ldots,x^{m+k}\right)$$

$$\in A^t\left(x^{m+k}\right) = \underset{i\in V}{\times}\, A_i^t\left(x^{m+k}_{\widetilde{N}(i)}\right)$$

holds. Therefore $\Delta^{(m)}$ is admissible according to Definition 3.18 **(1)**.

Having observed the independence property of $\Phi^{\Delta^\star}\left(x^0,\ldots,x^{m-1},x^m\right)$ from $\left(x^0,\ldots,x^{m-1}\right)$, for any fixed m, we now construct a sequence of policies

$$\Delta^\star_{(m)} = \left(\overset{(m)}{\Delta}{}^j : j\in\mathbb{N}\right), \qquad m\in\mathbb{N},$$

such that for fixed m, a similar independence holds. For $m=1$ we start with

$$\Delta^\star\left(x^0,x^1,\ldots,x^m,\ldots\right)$$

$$= \left(\overset{\star}{\Delta}{}^0\left(x^0\right),\overset{\star}{\Delta}{}^1\left(x^0,x^1\right),\ldots,\overset{\star}{\Delta}{}^m\left(x^0,x^1,\ldots,x^m\right),\ldots\right).$$

According to the construction performed above, we can find a control $\overset{\star}{\Delta}_{(1)}$ with

$$\overset{\star}{\Delta}_{(1)}\left(x^0,x^1,\ldots,x^m,\ldots\right)$$

$$= \left(\overset{(1)}{\Delta}{}^0\left(x^0\right),\overset{(1)}{\Delta}{}^1\left(x^1\right),\overset{(1)}{\Delta}{}^2\left(x^1,x^2\right),\ldots,\overset{(1)}{\Delta}{}^m\left(x^1,x^2,\ldots,x^m\right),\ldots\right),$$

which yields the same conditional expectation

$$\Phi^{\overset{\star}{\Delta}_{(1)}}(x^0) = \Phi^{\Delta^\star}(x^0)$$

as the optimal control $\Delta^\star$. We further have $\overset{(1)}{\Delta}{}^0(x^0) = \overset{\star}{\Delta}{}^0(x^0)$.

Continuing in the same way for $m = 2$, we apply the same procedure to strategy $\overset{\star}{\Delta}_{(1)}$ and obtain a new strategy

$$\overset{\star}{\Delta}_{(2)}\left(x^0, x^1, \ldots, x^m, \ldots\right)$$
$$= \left(\overset{(2)}{\Delta}{}^0(x^0), \overset{(2)}{\Delta}{}^1(x^1), \overset{(2)}{\Delta}{}^2(x^2), \overset{(2)}{\Delta}{}^2(x^2, x^3), \ldots\right.$$
$$\left. \ldots, \overset{(2)}{\Delta}{}^m(x^2, x^3, \ldots, x^m), \ldots\right),$$

and observe $\overset{(2)}{\Delta}{}^0(x^0) = \overset{(1)}{\Delta}{}^0(x^0)$, $\overset{(2)}{\Delta}{}^1(x^1) = \overset{(1)}{\Delta}{}^1(x^1)$.

We continue similarly and construct a sequence of strategies $\overset{\star}{\Delta}_{(n)}$, $n \in \mathbb{N}$, which is the starting point for the last step of our construction.

Following [VS64, p. 78], and using Weierstrass' diagonalization method we can construct a Markov strategy $\overset{\infty}{\Delta}(\beta) = \left\{\overset{\infty}{\Delta}_i(\beta), \; i \in V\right\}$ such that

$$\overset{\star}{\Delta}_{(\infty)}\left(x^0, x^1, \ldots, x^m, \ldots\right)$$
$$= \left(\overset{(\infty)}{\Delta}{}^0(x^0), \overset{(\infty)}{\Delta}{}^1(x^1), \overset{(\infty)}{\Delta}{}^2(x^2), \overset{(\infty)}{\Delta}{}^3(x^3), \ldots, \overset{(\infty)}{\Delta}{}^m(x^m), \ldots\right),$$

holds and additionally $\overset{(\infty)}{\Delta}{}^j(x^j) = \overset{(j)}{\Delta}{}^j(x^j)$, $\forall \, j \in \mathbb{N}$.

The construction further shows that the total expected discounted costs for all strategies $\overset{\star}{\Delta}_{(n)}$, $n \in \mathbb{N}$, are the same and equal that of the optimal strategy:

$$\mathfrak{R}_{x^0}^{\overset{\star}{\Delta}(n)}(\beta) = \mathfrak{R}_{x^0}^{\Delta^{\star}}(\beta), \qquad x_0 \in X.$$

From

$$\lim_{n \to \infty} \mathfrak{R}_{x^0}^{\overset{\star}{\Delta}(n)}(\beta) = \mathfrak{R}_{x^0}^{\overset{\infty}{\Delta}}(\beta), \qquad x_0 \in X,$$

(see [VS64, p. 78]) and because $\overset{\infty}{\Delta} \in LS_M$ we have

$$\mathfrak{R}_{x^0}^{\overset{\infty}{\Delta}}(\beta) = \mathfrak{R}_{x^0}^{\Delta^{\star}}(\beta). \qquad (3.17)$$

(c) Stationarity: We will show that it is possible to construct this Markov strategy (satisfying (3.17)) in the class LS_D of stationary Markov strategies. Bellman's equation for $\mathfrak{R}_x(\beta)$ is

$$\mathfrak{R}_{x^0}(\beta)$$

$$= \min_{\substack{\Delta^0 : \; \Delta^0(x^0)=a^0 \in A(x^0) \\ \Delta^0 \text{ local}}} \left\{ r(x^0, a^0) + \beta \sum_{y \in X} \mathfrak{R}_y(\beta) Q(y \mid x^0, a^0) \right\}. \qquad (3.18)$$

Because for the strategy $\overset{\infty}{\Delta}(\beta)$ equality (3.17) holds, determining decision $\overset{\infty}{\Delta}_i^0(x^0)$ is reduced to solving equation (3.18) in which $A(x^0) \subseteq A$. Suppose now that the decisions $\overset{\infty}{\Delta}_i^0(x^0), \ldots, \overset{\infty}{\Delta}_i^p(x^p)$, $i \in V$, are found and they depend on the respective observations $x^0, \ldots, x^p$ only. From inequality (3.14) the continuation

$$\overset{\infty}{\Delta}^{p+1}(\beta) = \left\{ \overset{\infty}{\Delta}_i^{p+1}(\beta) = \left\{ \overset{\infty}{\Delta}_i^{p+1}(x^{p+1}), \overset{\infty}{\Delta}_i^{p+2}(x^{p+2}), \ldots \right\}, \; i \in V \right\}$$

of strategy $\overset{\infty}{\Delta}(\beta)$ is such that

$$\mathsf{E}^{\overset{\infty}{\Delta}^p} \left\{ \sum_{t=p+1}^{\infty} \beta^t r\left(\xi^t, \overset{\infty}{\Delta}^t\right) \mid \xi^p = x^p \right\} \leq \mathsf{E}^{\Delta^p} \left\{ \sum_{t=p+1}^{\infty} \beta^t r\left(\xi^t, \Delta^t\right) \mid \xi^p = x^p \right\}$$

for each continuation

$$\Delta^{p+1}(\beta) = \left\{ \Delta_i^{p+1}(\beta) = \left(\Delta_i^{p+1}(x^{p+1}), \Delta_i^{p+2}(x^{p+2}), \ldots \right), \ i \in V \right\}$$

of the Markov strategy $\Delta(\beta)$. From the Markov property of strategy $\overset{\infty}{\Delta}(\beta)$ and (3.17) it follows that

$$\mathsf{E}^{\overset{\infty}{\Delta}_{x^p}^{p+1}} \left\{ \sum_{t=p}^{\infty} \beta^t r \left(\xi^t, \overset{\infty}{\Delta}^t \right) \mid \xi^p = x^p \right\} = \beta^{p+1} \mathfrak{R}_{x^p}(\beta).$$

Further, the optimal decisions $\overset{\infty}{\Delta}_i^p$, $i \in V$, for the strategy $\Delta(\beta)$ for every fixed x^p can be found as a solution of the equation

$$\mathfrak{R}_{x^p}(\beta) = \min_{\substack{\Delta^p: \ \Delta^p(x^p)=a^p \in A(x^p) \\ \Delta^p \ \text{local}}} \left\{ r(x^p, a^p) + \beta \sum_{y \in X} \mathfrak{R}_y(\beta) Q(y \mid x^p, a^p) \right\},$$

which coincides with Equation (3.18) because neither the decision space $A(x^p)$ nor the probabilities $Q(y \mid x^p, a^p)$ and costs $r(x^p, a^p)$ depend on time. Thus, we have proved that in the class of Markov strategies there exists some stationary strategy $\widetilde{\Delta}(\beta)$ for which

$$\mathfrak{R}_{x^0}^{\widetilde{\Delta}}(\beta) = \min_{\Delta \in LS_P} \mathfrak{R}_{x^0}^{\Delta}(\beta).$$

(II) Long time average costs: It is possible to find an infinite sequence of discount factors $\beta_\alpha \to 1$ as $\alpha \to \infty$ such that for all of them the optimal strategies $\widetilde{\Delta}(\beta_\alpha)$ coincide: $\widetilde{\Delta}(\beta_\alpha) = \Delta^\star$, $\Delta^\star = \left\{ \Delta_i^\star = \overset{\star}{\Delta}_i(x), i \in V \right\}$.

The existence of such sequence $\beta_\alpha \to 1$ as $\alpha \to \infty$ can be seen as follows. Take any sequence $\beta_{\alpha'} \to 1, \alpha' \to \infty$ with associated strategies $\widetilde{\Delta}(\beta_{\alpha'})$. All these strategies $\widetilde{\Delta}(\beta_{\alpha'}) = \{\widetilde{\Delta}_i(\beta_{\alpha'}), \ i \in V\}$ are by construction determined through the functions $\widetilde{\Delta}_i(\beta_{\alpha'}) = (\widetilde{\Delta}_i^t(\beta_{\alpha'}): t \in \mathbb{N})$,

$i \in V$, where $\left(\widetilde{\Delta}_i^t(\beta_{\alpha'}): t \in \mathbb{N}\right)$, are defined on finite sets $X_{\widetilde{N}(i)}$, $i \in V$, attaining therefore only a finite number of values.

So there exists an infinite subsequence $\beta_{\alpha'} \to 1$ of $\beta_\alpha \to 1$ which have the same associated strategy.

Let us consider this strategy $\Delta^\star$. The process ξ associated with $\Delta^\star$ is a homogeneous Markov chain (in fact, a time dependent controlled Markov random field according to Remark 3.30) with transition probabilities

$$\Pr\left\{x \mid y, \overset{\star}{\Delta}_i\left(y_{\widetilde{N}(i)}\right), \; i \in V\right\} =: \mathsf{Q}(x \mid y).$$

For this fixed strategy $\Delta^\star$ we consider

$$\limsup_{T \to \infty} \frac{1}{T+1} \sum_{t=0}^{T} \mathsf{E}_y^{\Delta^\star} \, r\left(\xi^t, \Delta^\star\right) = \rho(y, \Delta^\star).$$

Denoting $r(x) := r\left(x, \Delta^\star(x)\right)$ we obtain

$$\frac{1}{T+1} \sum_{t=0}^{T} \mathsf{E}_y^{\Delta^\star} \, r\left(\xi^t, \Delta^\star\right) = \sum_{x \in X} r(x)\varphi_{T,y}(x),$$

where

$$\varphi_{T,y}(x) := \frac{1}{T+1} \sum_{t=0}^{T} \Pr_y^{\Delta^\star} \left\{\xi^t = x\right\}.$$

Recalling the ergodic theorem for finite state Markov chains we obtain Cesaro convergence $\varphi_{T,y}(x)$ to a limiting probability

$$\lim_{T \to \infty} \varphi_{T,y}(x) =: \phi_y^\star(x),$$

which implies (similar to [VS64, p. 81]) for the fixed strategy $\Delta^\star$ the existence of the limit

$$\lim_{T \to \infty} \frac{1}{T+1} \sum_{t=0}^{T} \mathsf{E}_y^{\Delta^\star} \, r\left(\xi^t, \Delta^\star\right) = \sum_{x \in X} r(x)\phi_y^\star(x) = \rho(y, \Delta^\star).$$

Applying Abelian theorems this limit is obtained as well by the respective discounted costs for $\beta_\alpha \to 1$ [VS64, pp. 81, 82]. So the optimality of the stationary Markov strategy $\Delta^\star$ in the class LS_P of all local admissible strategies is proved.   □

Remark 3.40. We restricted our considerations to the study of nonrandomized strategies. Using Bellman's optimality equation and standard techniques (see, e.g., [Der70, Chapter 3, Theorem 1]), it can be shown that due to the finiteness of the decision space A for proving optimality considerations, it is not necessary to consider randomized strategies because some extremum will always be attained in the class of nonrandomized strategies.⊙

From Theorem 3.39 we conclude that determining an optimal strategy can be done using linear programming techniques; see [Der64] or [How60] for details.

Example 3.41 (Diffusion of knowledge and technologies). (Continuation of Example 3.23.) We sketch an example from the area of economical decision making; for details, see [DF93]. Let $\Gamma = (V, B)$ be a finite undirected graph without loops. The vertices represent organizations (or companies) and the edges indicate interactions between the organizations. Let $X_i = \{x_i^1, \ldots, x_i^{n_i}\}$ be the set of possible states for organization i, where the state x_i in X_i indicates the usage of a specific standard of technology by this organization. The global states of the group of organizations are $X = \underset{i \in V}{\times} X_i$.

According to [DF93] diffusion of knowledge and technologies can adequately be described by Markov random fields. This implies that local interactions are central to the description of the stochastic behavior of the systems over time as well.

Using the graph strcture underlying the spatial distribution of the firms, $\widetilde{N}(k)$ is the subgroup of organizations interacting with the k-th company and k itself. The locality property included in Definition 3.29 implies that decisions of organization k at time t depend only on the states and decisions of organizations interacting with the organization k and of those of k itself up to time t. Restricting decisions in a way that only local information is used is a natural assumption put on the behavior of many real systems.

In case that a company can only choose between two technologies at any time instant, we may describe this situation by a variant of the stochastic Ising model from statistical physics [Sin80, Rue69]. Transition probabilities for this model without controls are of the form [Kno96]

$$
\Pr\left\{\xi_j^{t+1} = x_j \mid \xi_{\widetilde{N}(j)}^t = y_{\widetilde{N}(j)}\right\}
$$

$$
= \frac{\exp\left[-\beta \sum_{k \in \widetilde{N}(j)} y_k x_j\right]}{\sum_{x_i \in \{+1,-1\}} \exp\left[-\beta \sum_{k \in \widetilde{N}(i)} y_k x_i\right]}, \quad (3.19)
$$

where y_k and x_j attain values ± 1 and we have $\beta > 0$.

Selection of the transition probability 3.19 is based on the experience that such modeling assumptions have been proven adequat in the area of some classes of locally interacting processes. β is in the models that originate from physics proportional to the inverse temperature. In the economic setting here it can be considered as a parametrization of the transition probabilities, which strengthens or weakens the tendency to change the actual state of the system.

In the spirit of our general models we want to determine an optimal strategy (under some of the possible cost functions) for subgroup $S \subset V$ of companies that is based on the previous behavior of companies in their neighborhood only. Then the set of admissible strategies when being in state x describes among others reaction on, say, investment decision, advertising, and organizational revisions, and possible additional restrictions on the admissible decisions by using $A(x) = \underset{i \in S}{\times} A_i\left(x_{\widetilde{N}(i)}\right)$, i.e., local information only. The expected average costs per time unit is a generally accepted criterion for evaluating production standards. From the results derived above, we conclude that there would exist an optimal stationary Markov nonrandomized strategy in the class of local strategies.

If we want to compute an optimal strategy (according to any optimality criterion) by solving a linear program (see Section 3.4.6), we usually explicitly need to know the stationary probabilities of the system. For the classical uncontrolled Ising model, the stationary probabilities may be found in [Lig85].

A first attempt to control the system's behavior is to consider constant controls $\Delta \equiv \beta$ where $\beta < \beta_c$, the critical temperature. But prescribing an "optimal" global Δ yields only an open loop control that has no local structure. The most natural local control for the transition kernel (3.19) would be to introduce in the spirit of our development *locally controlled temperature functions* $\beta_k(x_{\widetilde{N}(k)})$ for nodes $k \in V$. Then the model fits into the framework described above.   ⊙

3.4.4 Computational example: Cyclic queues

We reconsider the cyclic network of queues described in Example 3.32 and assume that the local decision makers at the nodes may adapt the service probabilities at their nodes according to the load of their own node and of the neighbored nodes. So their strategies are of local nature. According to Remark 3.40, optimal behavior of the local decision makers can be achieved already by using deterministic policies only.

The local decision spaces $A_i = \{a^1, a^2, \ldots, a^{n_i}\}$, $i = 1, \ldots, J$, are finite and do not depend on t. Then globally observed joint actions of the decision makers are from $A = \underset{i=1,\ldots,J}{\times} A_i$.

From Theorem 3.39 we know that an optimal local strategy $\Delta^\star$, (with respect to asymptotic average expected costs (3.10)) can be found in the class LS_D of deterministic stationary Markovian policies, i.e., $\Delta^\star = \left(\Delta_i^\star(x_{\widetilde{N}(i)}) : i = 1, 2, \ldots, J\right)$. We therefore consider in the following at time instant t for vertex j decision $\Delta_j^t(x_{\widetilde{N}(j)})$ according to Definition 3.18 (2), which depends on states of its neighborhood only. It follows that at the end of a time slot at node j, (if there are h customers present) a customer's service ends with probability $p_j(h,u) \in (0,1)$ and with probability $q_j(h,u) := 1 - p_j(h,u)$ that customer will stay there for at least one further time slot, $h \geq 1$, $u \in A_j$. Consequently, we obtain a local Markov strategy $\Delta = (\Delta_i\colon i = 1,2,\ldots,J)$ according to Definition 3.18 **(2)** and a controlled process (ξ, Δ), which is according to Definition 3.29 a controlled time dependent random field.

For any strategy $\Delta \in LS_D$ that results in service probabilities $p_j(h,u) \in (0,1)$ at the nodes, the controlled process (ξ, Δ) is ergodic on state space $S(K,J)$ with some stationary distribution $\pi^{K,J,\Delta} =: \pi^\Delta = (\pi^\Delta(x)\colon x \in S(K,J))$. But note that in general, (3.11) of Theorem 3.33 does not apply. The one-step costs occurring at time $t \in \mathbb{N}$ when being in state $\xi^t = x^t$ and applying action u^t are $r(x^t, u^t) \geq 0$. To determine an optimal strategy, we shall use the *strategy improvement procedure* [Der70, Section 6]:

Select some strategy Δ, and consider for some (unknown) function $v = (v(x)\colon x \in S(K,J))$ the equations [Der70, p. 56, 66]:

$$R_y^\Delta + v(y) = r(y, \Delta(y)) + \sum_{x \in S(K,J)} Q(x \mid y, \Delta(y))v(x),$$

$$y \in S(K,J) \quad (3.20)$$

and

$$\sum_{x \in S(K,J)} \pi^\Delta(x)v(x) = 0. \quad (3.21)$$

Solving for $\left\{ (v(x), R_y^\Delta)\colon x, y \in S(K,J) \right\}$ yields the costs R_y^Δ, when using Δ, which turns out to be independent of initial state y. The policy improvement algorithm is [Der70, p. 70]:

For each $y \in S(K,J)$, define A^y to be the set of actions a in state y for which

$$\sum_{x \in S(K,J)} Q(x \mid y, a)R_x^\Delta < R_y^\Delta,$$

or, if no actions satisfy the inequality, the set that satisfies

$$\sum_{x \in S(K,J)} Q(x \mid y, a) R_x^{\Delta} = R_y^{\Delta} \qquad (3.22)$$

and

$$r(y, a) + \sum_{x \in S(K,J)} Q(x \mid y, a) v^{\Delta}(x)$$

$$< r(y, \Delta(y)) + \sum_{x \in S(K,J)} Q(x \mid y, \Delta(y)) v^{\Delta}(x)$$

$$= R_y^{\Delta} + v^{\Delta}(y). \qquad (3.23)$$

Starting with the prescribed strategy Δ we define some local strategy $\Delta' \in LS_D$, which takes some action $a \in A^y$ in at least one state y for which A^y is nonempty; otherwise, the action taken is the one dictated by Δ.

Theorem 3.42 (see [Der70, p. 70, 71]). (a) If $\Delta' \neq \Delta$, then $R_y^{\Delta'} \leq R_y^{\Delta}, y \in S(K, J)$.

(b) The strategy improvement procedure leads to an optimal strategy within a finite number of iterations. If Δ' is the actual strategy and if A^y is empty for all y, then the actual policy is optimal: $\Delta' =: \Delta^\star$ and

$$R_x^{\Delta^\star} = \rho(x, \Delta^\star) = \inf_{\Delta \in LS_P} \rho(x, \Delta). \qquad \odot$$

To demonstrate the application of the procedure, we specialize now the data.

Example 3.43. Consider the cyclic network of queues described in Example 3.32 with node set $\{1, 2, 3\}$, and $K = 2$ customers. Selecting service probabilities is according to binary alternatives with $A_j \equiv \{0, 1\}$

$\forall \, j \in \{1, 2, 3\}$, where $a_j \in A_j$ indicates a decision for a specific service capacity. We assume $p_j(h, a)$ and $r(x, a)$ as

$$\begin{pmatrix} (p_1(1,0); p_1(1,1)) & (p_1(2,0); p_1(2,1)) \\ (p_2(1,0); p_2(1,1)) & (p_2(2,0); p_2(2,1)) \\ (p_3(1,0); p_3(1,1)) & (p_3(2,0); p_3(2,1)) \end{pmatrix} = \begin{pmatrix} (\frac{1}{3}; \frac{1}{2}) & (\frac{2}{3}; \frac{3}{4}) \\ (\frac{1}{4}; \frac{1}{2}) & (\frac{1}{3}; \frac{2}{3}) \\ (\frac{1}{4}; \frac{2}{3}) & (\frac{1}{2}; \frac{3}{4}) \end{pmatrix} \quad (3.24)$$

and

$$r(x, a) = \sum_{j \in S} \left[(c_j - a_j) x_j + a_j b_j \right], \quad (3.25)$$

where $c = (1, 2, 3)$, $b = (3, 1, 2)$, i.e., b_j is a cost associated to using additional service capacity for node j. We abbreviate the states of $S(3, 2)$ by

$$x^1 = (0, 0, 2), \qquad x^2 = (0, 1, 1), \qquad x^3 = (0, 2, 0),$$
$$x^4 = (1, 0, 1), \qquad x^5 = (1, 1, 0), \qquad x^6 = (2, 0, 0),$$

and the decision vectors

$$a^1 = (0, 0, 0), \quad a^2 = (0, 0, 1), \quad a^3 = (0, 1, 0), \quad a^4 = (0, 1, 1),$$
$$a^5 = (1, 0, 0), \quad a^6 = (1, 0, 1), \quad a^7 = (1, 1, 0), \quad a^8 = (1, 1, 1).$$

Then the transition kernels $Q(x \mid y, a)$ are given by

Q	$(0, 0, 2)$	$(0, 1, 1)$	$(0, 2, 0)$
x^1	$q_3(2, a_3)$	$p_2(1, a_2) q_3(1, a_3)$	0
x^2	0	$q_2(1, a_2) q_3(1, a_3)$	$p_2(2, a_2)$
x^3	0	0	$q_2(2, a_2)$
x^4	$p_3(2, a_3)$	$p_2(1, a_2) p_3(1, a_3)$	0
x^5	0	$q_2(1, a_2) p_3(1, a_3)$	0
x^6	0	0	0

Q	$(1,0,1)$	$(1,1,0)$	$(2,0,0)$
x^1	0	0	0
x^2	$p_1(1,a_1)q_3(1,a_3)$	$p_1(1,a_1)p_2(1,a_2)$	0
x^3	0	$p_1(1,a_1)q_2(1,a_2)$	0
x^4	$q_1(1,a_1)q_3(1,a_3)$	$q_1(1,a_1)p_2(1,a_2)$	0
x^5	$p_1(1,a_1)p_3(1,a_3)$	$q_1(1,a_1)q_2(1,a_2)$	$p_1(2,a_1)$
x^6	$q_1(1,a_1)p_3(1,a_3)$	0	$q_1(2,a_1)$

We start with strategy $\Delta(x) \equiv a^1 = (0,0,0)$. From (3.24)–(3.25) we have for this Δ: $p_j(x_j) = p_j\big(x_j, \Delta(x_j)\big)$ and $r(x) = r\big(x, \Delta(x)\big)$ as follows:

$$\begin{pmatrix} p_1(1) & p_1(2) \\ p_2(1) & p_2(2) \\ p_3(1) & p_3(2) \end{pmatrix} = \begin{pmatrix} 1/3 & 2/3 \\ 1/4 & 1/3 \\ 1/4 & 1/2 \end{pmatrix}$$

$r(x^1)$	$r(x^2)$	$r(x^3)$	$r(x^4)$	$r(x^5)$	$r(x^6)$
6	5	4	4	3	2

For the prescribed Δ, Theorem 3.33 applies and we obtain the ergodic distribution $\pi^\Delta = \big(\pi^\Delta(x)\colon x \in S(K,J)\big)$ of ξ and costs R^Δ:

$$\pi^\Delta = \left(\frac{3}{29}; \frac{8}{29}; \frac{9}{58}; \frac{6}{29}; \frac{6}{29}; \frac{3}{58} \right),$$

and

$$R^\Delta = \sum_{x \in S(K,J)} \pi^\Delta(x) r(x) = \frac{121}{29}.$$

Applying (3.20)–(3.21) we have $v^\Delta(x)$:

$v^\Delta(x^1)$	$v^\Delta(x^2)$	$v^\Delta(x^3)$	$v^\Delta(x^4)$	$v^\Delta(x^5)$	$v^\Delta(x^6)$
$\dfrac{7469}{3364}$	$\dfrac{31023}{16820}$	$\dfrac{22323}{16820}$	$-\dfrac{4827}{3364}$	$-\dfrac{31153}{16820}$	$\dfrac{85963}{16820}$

We determine the sets A^y, $y \in S(K, J)$ at this stage of the algorithm. Equation (3.22) is satisfied for each $y \in S(K, J)$. From (3.23) we have that $A^{x^1} \ni a^2$, $A^{x^2} \ni a^2$, $A^{x^3} \ni a^3$, $A^{x^4} \ni a^2$, $A^{x^5} = \emptyset$, $A^{x^6} = \emptyset$.

Thus, we may define a new strategy Δ' as

$\Delta'(x^1)$	$\Delta'(x^2)$	$\Delta'(x^3)$	$\Delta'(x^4)$	$\Delta'(x^5)$	$\Delta'(x^6)$
a^2	a^2	a^3	a^2	a^1	a^1

or shortly, $\Delta'(x) = (0, x_2 \text{ div } 2, \text{sign } x_3)$, where a div b is the integral part obtained from dividing a by b, sign a is the sign of number a (sign $a \in \{-1; 0; 1\}$).

For the new strategy $\Delta'(x)$ (3.24)–(3.25) yield $p'_j(x_j) = p_j(x_j, \Delta'(x_j))$ and $r'(x) = r(x, \Delta(x))$:

$$\begin{pmatrix} p'_1(1) & p'_1(2) \\ p'_2(1) & p'_2(2) \\ p'_3(1) & p'_3(2) \end{pmatrix} = \begin{pmatrix} 1/3 & 2/3 \\ 1/4 & 2/3 \\ 2/3 & 3/4 \end{pmatrix}$$

$r(x^1)$	$r(x^2)$	$r(x^3)$	$r(x^4)$	$r(x^5)$	$r(x^6)$
6	6	3	5	3	2

Theorem 3.33 yields the ergodic distribution $\pi^{\Delta'} = (\pi^{\Delta'}(x) \colon x \in S(K, J))$ and costs $R^{\Delta'}$:

$$\pi^{\Delta'} = \left(\frac{1}{46}; \frac{9}{46}; \frac{27}{184}; \frac{27}{184}; \frac{9}{23}; \frac{9}{92} \right),$$

and

$$R^{\Delta'} = \sum_{x \in S(K,J)} \pi^{\Delta'}(x) r'(x) = \frac{177}{46}.$$

Solving (3.20)–(3.21) for $v^{\Delta'}(x)$ we have:

$v^{\Delta'}(x^1)$	$v^{\Delta'}(x^2)$	$v^{\Delta'}(x^3)$	$v^{\Delta'}(x^4)$	$v^{\Delta'}(x^5)$	$v^{\Delta'}(x^6)$
$\dfrac{103227}{42320}$	$\dfrac{212523}{84640}$	$\dfrac{104883}{84640}$	$-\dfrac{36429}{84640}$	$-\dfrac{67801}{84640}$	$-\dfrac{302401}{84640}$

Equation (3.22) is now satisfied for all $y \in S(K, J)$. From (3.23) we conclude that $A^y = \emptyset$ for all $y \in S(K, J)$. So the strategy $\Delta^\star = (\Delta'(x) = (0, x_2 \text{ div } 2, \text{sign } x_3), x \in S(K, J))$ is optimal and $R^{\Delta^\star} = \frac{177}{46}$.   ⊙

3.4.5 Separable cost functions and global optimality of local strategies

Up to now we were mainly concerned with finding optimal strategies in classes of local strategies. Clearly this is an essential problem if (as is often the case) local strategies are the only possible policies that can be applied because of practical limitations.

But there is a related problem that deals with the question whether local strategies can be globally optimal. We find in this section conditions that imply that locally defined policy classes contain strategies that are optimal even in the class of all admissible policies.

For now we consider cost functions $r(x, u)$ that are separable (additive) functions with respect to the neighborhood graph as follows.

Definition 3.44. A cost function

$$r \colon \bar{\kappa} = \{(x, a) \colon x \in X,\ a \in A(x)\} \subset X \times A \to \mathbb{R}$$

is separable with respect to the neighborhood structure $\Gamma = (V, B)$ if it is of the form

$$r(x, a) = \sum_{i \in V} r_i\left(x_{\widetilde{N}(i)}, a_i\right),$$

i.e., if it is a sum of locally defined measurable bounded reward functions $r_i\left(x_{\widetilde{N}(i)}, a_i\right)$.   ⊙

In this section we assume that according to Definition 2.16, the control sets are of product form and may be globally dependent on states but are independent of time.

Assumption 3.45. For any system configuration (state) x^t and any vertex $i \in V$, let $A_i(x^t)$ be the set of admissible actions (admissible decision values) for decisions at node i at time t if $\xi^t = x^t$. We denote $A(x^t) = \underset{i \in V}{\times} A_i(x^t)$. $\odot$

Remark 3.46. A (randomized) strategy π is admissible according to Definition 2.17 if for every $t \in \mathbb{N}$

$$\pi^t(x^0, \ldots, x^{t-1}, x^t) = \left\{ \pi_i^t(x^0, \ldots, x^{t-1}, x^t) : i \in V \right\} \in A(x^t).$$

A strategy $\pi = (\pi_i : i \in V)$ is Markovian if for all $x^0, x^1, \ldots, x^t \in X$ the local decision rules depend only on the present global state x^t: $\pi_i^t(x^0, \ldots, x^t) = \pi_i^t(x^t)$, $i \in V$. $\odot$

Theorem 3.47. Suppose there exist a constant g and a Borel function $v(x)$ on X such that for the transition function Q (Definition 3.29) of the controlled process with locally interacting synchronous components holds

$$g + v(x) = \min_{a \in A(x)} \left\{ r(x, a) + \sum_{y \in X} v(y) Q(y \mid x, a) \right\}, \qquad x \in X.$$

Then

$$\inf_{\pi \in LS} \rho(x, \pi) \geq g, \qquad x \in X.$$

Moreover, if for a strategy $\Delta^\star \in LS_D$ holds

$$g + v(x) = r(x, \Delta^\star(x)) + \sum_{y \in X} v(y) Q(y \mid x, \Delta^\star(x)), \qquad x \in X,$$

then $\Delta^{\star}$ is optimal and for all $x \in X$

$$\rho(x, \Delta^{\star}) \equiv g. \qquad \qquad \odot$$

Proof: It is obvious that for any strategy $\pi \in LS$ we have

$$\mathsf{E}_{x^0}^{\pi}\left\{\sum_{t=1}^{T}\left[v(\xi^t) - \mathsf{E}^{\pi}\left\{v(\xi^t) \mid \xi^0 = x^0, \alpha^0 = a^0, \ldots\right.\right.\right.$$

$$\left.\left.\left.\ldots, \xi^{t-1} = x^{t-1}, \alpha^{t-1} = a^{t-1}\right\}\right]\right\} = 0.$$

From the definition and from the Markov property follows

$$\mathsf{E}^{\pi}\left\{v(\xi^t) \mid \xi^0 = x^0, \alpha^0 = a^0, \ldots, \xi^{t-1} = x^{t-1}, \alpha^{t-1} = a^{t-1}\right\}$$

$$= \sum_{y \in X} v(y)\mathsf{Q}(y \mid x^{t-1}, a^{t-1})$$

$$= r(x^{t-1}, a^{t-1}) + \sum_{y \in X} v(y)\mathsf{Q}(y \mid x^{t-1}, a^{t-1}) - r(x^{t-1}, a^{t-1})$$

$$\geq g + v(x^{t-1}) - r(x^{t-1}, a^{t-1}).$$

Therefore

$$\mathsf{E}_{x^0}^{\pi}\sum_{t=1}^{T}\left[r(\xi^{t-1}, \alpha^{t-1}) + v(\xi^t) - v(\xi^{t-1}) - g\right] \geq 0, \qquad (3.26)$$

and consequently

$$\mathsf{E}_{x^0}^{\pi}\left\{\sum_{t=1}^{T} r(\xi^{t-1}, \alpha^{t-1})\right\} + v(\xi^T) - v(x^0) \geq Tg. \qquad (3.27)$$

Because X is finite and therefore $v(x)$ is bounded, from (3.27) we obtain

$$\rho(x, \pi) \geq g, \qquad x \in X.$$

The second part of the theorem follows from the fact that for strategy $\Delta^\star$ in (3.26) and (3.27), equality holds. $\square$

In general, local strategies cannot be expected to be globally optimal. But under mild technical assumptions, we can prove such a statement. We assume henceforth in this section the following.

Assumption 3.48. There exist non-negative measures μ_i on X_i such that $\forall\, i \in V$ we have $\mu_i(X_i) > 0$ and such that for all $i \in V$:

$$\mu_i(x_i) \leq \mathsf{Q}\Big(x_i \mid y_{\widetilde{N}(i)}, a_i\Big), \text{ for } x_i \in X_i, y_{\widetilde{N}(i)} \in X_{\widetilde{N}(i)}, a_i \in A_i. \qquad \odot$$

Let $S(X)$ denote the Banach space of (bounded) functions $v(x) = \sum_{i\in V} v_i(x_i)$ on $X = \times_{i\in V} X_i$, with norm $\|v\| = \sum_{i\in V} \max_{x_i\in X_i} |v_i(x_i)|$. Denote the operator $\mathfrak{U}$ on $S(X)$ by:

$$\mathfrak{U}v(x) = \min_{a\in A(x)} \left\{ \sum_{i\in V} r_i\Big(x_{\widetilde{N}(i)}, a_i\Big) + \sum_{y\in X} v(y)\mathsf{Q}'(y \mid x, a) \right\},$$

where

$$\mathsf{Q}'(y \mid x, a) = \mathsf{Q}(y \mid x, a) - \prod_{i\in V} \mu_i(y_i).$$

Denote the set of minimizers for $\mathfrak{U}v(x)$

$$A'_v(x) = \left\{ a \in A(x) \colon \mathfrak{U}v(x) = \sum_{i\in V} r_i\Big(x_{\widetilde{N}(i)}, a_i\Big) + \sum_{y\in X} v(y)\mathsf{Q}'(y \mid x, a) \right\}$$

and define the local operators

$$A_i^{v_i}\Big(x_{\widetilde{N}(i)}, a_i\Big) = r_i\Big(x_{\widetilde{N}(i)}, a_i\Big) +$$

$$+ \sum_{y_i\in X_i} v_i(y_i) \left[\mathsf{Q}_i\Big(y_i \mid x_{\widetilde{N}(i)}, a_i\Big) - \mu_i(y_i) \prod_{j\colon \left\{\substack{j\in V;\\ j\neq i.}\right.} \mu_j(X_j) \right].$$

Theorem 3.49. Under Assumption 3.48, in class LS_D a globally optimal strategy exists. ⊙

Proof: Using standard arguments from [GS72b], it follows that the operator $\mathfrak{U}$ is a contraction on $S(X)$. By Banach's fixed point theorem in $S(X)$ exists a fixed point $v^\star$ such that

$$\mathfrak{U}v^\star(x) = v^\star,$$

which is

$$v^\star(x) = \min_{a \in A(x)} \left\{ \sum_{i \in V} r_i\left(x_{\widetilde{N}(i)}, a_i\right) + \sum_{y \in X} v^\star(y) Q'(y \mid x, a) \right\}, \qquad x \in X.$$

Using the structure of the operator $\mathfrak{U}$ we have

$$\mathfrak{U}v^\star(x) = \sum_{i \in V} \min_{a_i \in A_i} A_i^{v_i^\star}\left(x_{\widetilde{N}(i)}, a_i\right).$$

Because the A_i are finite, there exists measurable functions $\Delta_i^\star = \Delta_i^\star\left(x_{\widetilde{N}(i)}\right)$ such that

$$\min_{a_i \in A_i(x)} A_i^{v_i^\star}\left(x_{\widetilde{N}(i)}, a_i\right) = A_i^{v_i^\star}\left(x_{\widetilde{N}(i)}, \Delta_i^\star\left(x_{\widetilde{N}(i)}\right)\right).$$

Therefore there exists a function $\Delta^\star = \left\{ \Delta_i^\star\left(x_{\widetilde{N}(i)}\right) : i \in V \right\} \in LS_D$ such that

$$\mathfrak{U}v^\star(x) = \sum_{i \in V} r_i\left(x_{\widetilde{N}(i)}, \Delta_i^\star\left(x_{\widetilde{N}(i)}\right)\right) + \sum_{y \in X} v^\star(y) Q'(y \mid x, \Delta^\star(x))$$

and

$$v^\star(x) = \sum_{i \in V} r_i\left(x_{\widetilde{N}(i)}, \Delta_i^\star\left(x_{\widetilde{N}(i)}\right)\right) + \sum_{y \in X} v^\star(y) Q'(y \mid x, \Delta^\star(x))$$

or

$$g + v^\star(x) = \sum_{i \in V} r_i \left(x_{\widetilde{N}(i)}, \Delta_i^\star \left(x_{\widetilde{N}(i)} \right) \right) + \sum_{y \in X} v^\star(y) \mathsf{Q}(y \mid x, \Delta^\star(x)),$$

where

$$g = \sum_{i \in V} \sum_{y \in X} v_i^\star(y_i) \mu_i(y_i) \prod_{j:\ \left\{ \begin{smallmatrix} j \in V; \\ j \neq i. \end{smallmatrix} \right.} \mu_j(X_j).$$

Therefore the conditions of the Theorem 3.47 are satisfied and strategy $\Delta^\star \in LS_D$ is optimal. $\qquad\qquad\qquad\qquad\qquad\qquad\qquad\qquad$ □

3.4.6   Computing optimal strategies with linear programs

In this subsection we consider a controlled process (ξ, α) with finite state and action space. We reduce the problem of finding the optimal control to the problem of solving a linear program. For more details on the general principle, see [Der64, Der70].

As before, we assume that for $K \subset V$, $y \in X$, $a \in A(y)$ holds

$$\Pr \left\{ \xi_K^{t+1} = x_K \mid \xi^0 = x^0, \alpha^0(\xi^0) = a^0, \ldots, \xi^t = y, \alpha^t(\xi^0, \ldots, \xi^t) = a \right\}$$

$$= \prod_{j \in K} \Pr \left\{ \xi_j^{t+1} = x_j \mid \xi_{\widetilde{N}(j)}^t = y_{\widetilde{N}(j)}, \alpha_j^t \left(\xi_{\widetilde{N}(j)}^0, \ldots, \xi_{\widetilde{N}(j)}^t \right) = a_j \right\}$$

$$= \prod_{j \in K} \mathsf{Q}_j \left(x_j \mid y_{\widetilde{N}(j)}, a_j \right) = \mathsf{Q}_K(x_K \mid y, a).$$

If $K = V$, then $\mathsf{Q}(x \mid y, a) = \Pr \left\{ \xi^{t+1} = x \mid \xi^t = y, \alpha^t = a \right\}$, with $\sum_{x \in X} \mathsf{Q}(x \mid y, a) = 1$ for $y \in X$.

We further assume that the strategies π that are applied are stationary Markov, but not necessarily local, i.e., $\pi \in \Pi_S$. We have (Definition 2.18) for all $t \in \mathbb{N}$, $y \in X$, $a \in A$,

$$\Pr \left\{ \alpha^t = a \mid \xi^t = y \right\} = \pi(y; a).$$

It follows that ξ is for any fixed policy π a time homogeneous Markov chain with transition probability

$$p^\pi(y; x) = \Pr\left\{\xi^{t+1} = x \mid \xi^t = y\right\}$$
$$= \sum_{a \in A} \pi(y; a) Q(x \mid y, a), \quad x, y \in X.$$

We further assume that for any strategy π, the transition kernel $p^\pi(y; x)$ of the resulting Markov chain ξ is ergodic with the same unique positive recurrent class X. Then under any strategy π there is a unique limiting and stationary distribution $p^\pi = (p^\pi(x): x \in X)$ for ξ, which may depend on π.

The cost function is assumed to be separable according to Definition 3.44 with respect to the underlying neighborhood structure.

The expected time averaged costs under policy π with initial state $y \in X$ up to time T are

$$\mathsf{E}_y^\pi \frac{1}{T+1} \sum_{t=0}^{T} r(\xi^t, \alpha^t) = \sum_{x \in X} \sum_{a \in A} r(x, a) \frac{1}{T+1} \sum_{t=0}^{T} \Pr_y^\pi(\xi^t = x, \alpha^t = a).$$

From ergodicity of ξ for fixed π it follows

$$\frac{1}{T+1} \sum_{t=0}^{T} \Pr_y^\pi(\xi^t = x, \alpha^t = a) \xrightarrow{T \to \infty} p^\pi(x) \cdot \pi(x; a),$$

independent of the initial state $y \in X$. Therefore the asymptotic expected average costs under π with initial state $y \in X$ are

$$\rho(y, \pi) = \limsup_{T \to \infty} \mathsf{E}_y^\pi \frac{1}{T+1} \sum_{t=0}^{T} r(\xi^t, \alpha^t)$$
$$= \sum_{x \in X} \sum_{a \in A} r(x, a) p^\pi(x) \cdot \pi(x; a).$$

Abbreviating

$$z^\pi(x, a) = p^\pi(x) \cdot \pi(x; a)$$

our problem is now (independent of the initial state $y \in X$) to solve the problem

$$\min_{\pi \in \Pi_S} \sum_{x \in X} \sum_{a \in A} r(x, a) z^\pi (x, a).$$

We therefore consider the following linear program:
Find

$$\min_{(z(x,a)\,:\, x \in X, a \in A)} \sum_{x \in X} \sum_{a \in A} r(x, a) z(x, a)$$

subject to the constraints

$$z(x, a) \geq 0, \qquad x \in X, \ a \in A$$

$$\sum_{a \in A} z(x, a) = \sum_{y \in X} \sum_{a \in A} z(y, a) Q(x \mid y, a), \qquad x \in X,$$

$$\sum_{y \in X} \sum_{a \in A} z(y, a) = 1.$$

Any solution of this problem has the property [Der63, p. 21]

$$\sum_{a \in A} z(y, a) > 0 \quad \text{for all } y \in X,$$

and setting

$$\pi^\star(x; a) = \frac{z(x, a)}{\sum_{a \in A} z(x, a)} \quad \text{for all } x \in X, \ a \in A,$$

yields an optimal policy $\pi^\star \in \Pi_S$.

   A further step of the solution procedure is to find an optimal deterministic (stationary) policy. This can be done simply by solving the linear program with the simplex algorithm. This leads to a solution that is an extremal point of the convex polyeder of the admissible solutions, which is known to represent a deterministic policy. From Theorem 3.49 we know that due to the assumption of having a separable cost function, there exists a local stationary deterministic policy that is optimal in the class of all stationary Markov policies.

The obtained deterministic policy, which is not necessarily local, can be used to then find an optimal or nearly optimal local policy. But at present, there seems to be no general algorithm known to perform this procedure.

3.5 LOCAL CONTROL OF INTERACTING MARKOV PROCESSES ON GRAPHS WITH COMPACT STATE AND ACTION SPACES

In this section we search for optimal policies with respect to the (maximin) reward criterion of Definition 2.22, which maximizes the minimal reward $\phi(y, \pi)$ for given policy π and starting in y. Compactness and continuity conditions as imposed here on state space and action space are standard assumptions for many investigations in the literature on stochastic dynamic optimization or dynamic programming; see [Sch75, Bal89] and the references there. Our new point of view is the local structure imposed on these spaces.

3.5.1 Existence of optimal Markov strategies in the class of local strategies

In this subsection we prove statements on the existence of best policies in subsets of the class of all policies for a given problem setting extending [DKC01]). Following our general path of investigations, the statements of the theorems are concerned with the case of local policies for systems driven by local and synchronous kernels (see Definition 3.21, but we point out that the proofs work for general classes as well). So our first statements are without reference to locally determined processes. Clearly optimal policies in a subclass of all policies may be globally suboptimal, although the local optimum is attained within the prescribed class as will be shown.

Throughout this subsection the following assumption is in force.

Assumption 3.50. The admissible control sets are independent of the time of decision:

$$A(x) := A^t(x) = \underset{i \in V}{\times} A_i^t\left(x_{\widetilde{N}(i)}\right) =: \underset{i \in V}{\times} A_i\left(x_{\widetilde{N}(i)}\right), \quad \forall\, t \in \mathbb{N}. \quad \odot$$

The following preparatory result was given for the Markov framework in [GS72a] without proof.

Theorem 3.51. If there exist bounded Borel functions $g(x)$ and $v(x)$ on X such that for all $t \in \mathbb{N}$ and admissible control sets $A(x) := \underset{i \in V}{\times} A_i\left(x_{\widetilde{N}(i)}\right)$

$$g(x) \geq \sup_{a \in A(x)} \int g(y) Q(dy \mid x, a), \quad x \in X, \tag{3.28}$$

$$g(x) + v(x) \geq \sup_{a \in A(x)} \left\{ r(x, a) + \int v(y) Q(dy \mid x, a) \right\}, \quad x \in X, \tag{3.29}$$

then

$$\sup_{\pi \in LS} \phi(x, \pi) \leq g(x), \quad x \in X. \tag{3.30}$$

If additionally for a strategy $\pi^* \in LS$

$$g(x) = \int g(y) Q\left(dy \mid x, \pi^*(x)\right), \quad x \in X,$$

$$g(x) + v(x) = r\left(x, \pi^*(x)\right) + \int v(y) Q\left(dy \mid x, \pi^*(x)\right), \quad x \in X,$$

holds, then $\phi(x, \pi^*) \equiv g(x)$, and π^* is optimal in LS (locally optimal). $\odot$

Proof: For any strategy $\pi \in LS$

$$\mathsf{E}^\pi_{x^0}\left\{\sum_{t=1}^T\left[v(\xi^t) - \mathsf{E}^\pi\left\{v(\xi^t) \mid \xi^0 = x^0, \alpha^0 = a^0, \ldots \right.\right.\right.$$

$$\left.\left.\left. \ldots, \xi^{t-1} = x^{t-1}, \alpha^{t-1} = a^{t-1}\right\}\right]\right\} = 0. \quad (3.31)$$

From (3.9) and (3.29) we have

$$\mathsf{E}^\pi\left\{v(\xi^t) \mid \xi^0 = x^0, \alpha^0 = a^0, \ldots, \xi^{t-1} = x^{t-1}, \alpha^{t-1} = a^{t-1}\right\}$$

$$= \int v(y)\mathsf{Q}(dy \mid x^{t-1}, a^{t-1})$$

$$= r(x^{t-1}, a^{t-1}) + \int v(y)\mathsf{Q}(dy \mid x^{t-1}, a^{t-1}) - r(x^{t-1}, a^{t-1})$$

$$\leq g(x^{t-1}) + v(x^{t-1}) - r(x^{t-1}, a^{t-1}). \quad (3.32)$$

Inserting (3.32) into (3.31), we obtain

$$\mathsf{E}^\pi_{x^0}\left\{\sum_{t=1}^T r(\xi^{t-1}, a^{t-1}) + v(\xi^T) - v(x^0)\right\} \leq \mathsf{E}^\pi_{x^0}\sum_{t=1}^T g(\xi^{t-1}). \quad (3.33)$$

From (3.28) we have

$$\mathsf{E}^\pi_{x^0}g(\xi^t)$$

$$= \mathsf{E}^\pi_{x^0}\left\{\mathsf{E}^\pi\left[g(\xi^t) \mid \xi^0 = x^0, \alpha^0 = a^0, \ldots, \xi^{t-1} = x^{t-1}, \alpha^{t-1} = a^{t-1}\right]\right\}$$

$$= \mathsf{E}^\pi_{x^0}\left\{\int g(y)\mathsf{Q}(dy \mid \xi^{t-1}, \alpha^{t-1})\right\}$$

$$\leq \mathsf{E}^\pi_{x^0}g(\xi^{t-1}) \leq \cdots \leq \mathsf{E}^\pi_{x^0}g(\xi^0)$$

$$= g(x^0), \quad (3.34)$$

and, consequently,

$$\mathsf{E}^\pi_{x^0}\sum_{t=1}^T g(\xi^{t-1}) \leq Tg(x^0). \quad (3.35)$$

Inserting (3.35) in (3.33), we obtain from the boundedness of $g(x)$ and $v(x)$ inequality (3.30).

To prove of the second part of the theorem, we note that for the strategy $\pi^\star \in LS$ as described the inequalities (3.32)–(3.35) are equalities. □

Corollary 3.52. If (3.28) and (3.29) hold and a strategy $\pi^\star \in LS_S \subseteq LS$ fulfills

$$g(x) = \int g(y)Q(dy \mid x, \pi^\star(x)), \qquad x \in X,$$

$$g(x) + v(x) = r(x, \pi^\star(x)) + \int v(y)Q(dy \mid x, \pi^\star(x)), \qquad x \in X,$$

then there exist admissible deterministic local policies that are optimal in LS (locally optimal). ⊙

From Theorem 3.51 we directly obtain the following criterion.

Theorem 3.53. Assume there exist a constant g and a bounded Borel function $v(x)$ on X such that

$$g + v(x) = \sup_{a \in A(x)} \left\{ r(x, a) + \int v(y)Q(dy \mid x, a) \right\}, \qquad x \in X. \quad (3.36)$$

Then

$$\sup_{\pi \in LS} \phi(x, \pi) \le g, \qquad x \in X.$$

Moreover, if for a strategy $\pi^\star \in LS$

$$g + v(x) = r(x, \pi^\star(x)) + \int v(y)Q(dy \mid x, \pi^\star(x)), \qquad x \in X,$$

then $\phi(x, \pi^\star) \equiv g$ and $\pi^\star$ is optimal in LS. ⊙

Remark 3.54. Corollary 3.52 applies here again.   ⊙

The following regularity assumptions on the transition kernels provide conditions for proving optimality properties of local strategies. Recall the definition

$$\bar{\kappa} = \big\{(x, a) \in X \times A \text{ with } a \in A(x)\big\}$$

and Assumption 3.50:

$$A(x) := A^t(x) = \underset{i \in V}{\times} A_i^t\big(x_{\tilde{N}(i)}\big) =: \underset{i \in V}{\times} A_i\big(x_{\tilde{N}(i)}\big).$$

Assumption 3.55. $Q = \prod_{j \in V} Q_j$ is according to Definition 3.10 a local and synchronous Markov kernel for (ξ, π).

There exists a non-negative measure μ on $(X, \mathfrak{X})$ with $\mu(X) > 0$ such that:

$$\mu(C) \leq Q(C \mid x, a), \qquad C \in \mathfrak{X}, \quad (x, a) \in \bar{\kappa}. \qquad \qquad ⊙$$

Assumption 3.56. $Q = \prod_{j \in V} Q_j$ is according to Definition 3.10 a local and synchronous Markov kernel for (ξ, π).

There exists a non-negative measure η on $(X, \mathfrak{X})$ with $\eta(X) < 2$ such that:

$$\eta(C) \geq Q(C \mid x, a), \qquad C \in \mathfrak{X}, \quad (x, a) \in \bar{\kappa}. \qquad \qquad ⊙$$

Let $M(X)$ be the Banach space of bounded Borel measurable functions on X with norm $\|u\| = \sup_{x \in X} |u(x)|$. We define operators $\mathfrak{U}'$ and $\mathfrak{U}''$ on $u \in M(X)$ by:

$$\mathfrak{U}'u(x) = \sup_{a \in A(x)} \left\{ r(x, a) + \int u(y) Q'(dy \mid x, a) \right\},$$

$$\mathfrak{U}''u(x) = -\sup_{a \in A(x)} \left\{ r(x, a) + \int u(y) Q''(dy \mid x, a) \right\},$$

where

$$Q'(C \mid x, a) = Q(C \mid x, a) - \mu(C),$$
$$Q''(C \mid x, a) = \eta(C) - Q(C \mid x, a).$$

For every function $u \in M(X)$ we denote

$$A_u'(x) = \left\{ a \colon a \in A(x),\ \mathfrak{A}'u(x) = r(x, a) + \int u(y)Q'(dy \mid x, a) \right\},$$
$$A_u''(x) = \left\{ a \colon a \in A(x),\ -\mathfrak{A}''u(x) = r(x, a) + \int u(y)Q''(dy \mid x, a) \right\}.$$

Using these definitions, we define maps

$$A_u' \colon X \to 2^A, \qquad x \to A_u'(x)$$

and

$$A_u'' \colon X \to 2^A, \qquad x \to A_u''(x).$$

Theorem 3.57. Let Assumption 3.55 (Assumption 3.56) hold and assume further:

1) The operator $\mathfrak{A}'$ (operator $\mathfrak{A}''$) transforms some metric subspace $S(X) \subseteq M(X)$ (with metric ρ induced by the norm of $M(X)$) into itself;

2) The map A_u' (map A_u') is measurable (Definition 2.34) for every function $u \in S(X)$.

Then there exists a strategy in LS_S that is optimal in LS (locally optimal). ⊙

The proof is similar to [GS72b, Theorems 2 and 2'] and therefore we only sketch the main ideas:

The operator $\mathfrak{A}'$ (operator $\mathfrak{A}''$) is a contraction on $S(X) \subseteq M(X)$, so the equation $\mathfrak{A}'u = u$ ($\mathfrak{A}''u = u$) has a unique solution $u^\star(x)$ in $S(X)$.

Explicitly this is

$$u^*(x) = \sup_{a \in A(x)} \left\{ r(x,a) + \int u^*(y) Q'(dy \mid x,a) \right\}, \qquad (3.37)$$

respectively

$$u^*(x) = - \sup_{a \in A(x)} \left\{ r(x,a) + \int u^*(y) Q''(dy \mid x,a) \right\}. \qquad (3.38)$$

(3.37) ((3.38), respectively) are the Optimality Equation (3.36) with $v = u^*$ ($v = -u^*$) and $g = \int u^*(x)\,\mu(dx)$ ($g = \int u^*(x)\,\eta(dx)$). Therefore there exists $a^* \in A'_{u^*}(x)$ ($a^{**} \in A''_{u^*}(x)$) such that

$$\sup_{a \in A(x)} \left\{ r(x,a) + \int u^*(y) Q'(dy \mid x,a) \right\}$$

$$= r(x,a^*) + \int u^*(y) Q'(dy \mid x,a^*)$$

$$\left(\sup_{a \in A(x)} \left\{ r(x,a) + \int u^*(y) Q''(dy \mid x,a) \right\} \right.$$

$$\left. = r(x,a^{**}) + \int u^*(y) Q''(dy \mid x,a^{**}) \right).$$

From measurability by the Theorem of Choice, the map A'_u (map A''_u) has a Borel measurable selector $\pi^* : X \to A$ ($\pi^{**} : X \to A$) such that

$$r(x,a^*) + \int u^*(y) Q'(dy \mid x,a^*)$$

$$= r\big(x,\pi^*(x)\big) + \int u^*(y) Q'\big(dy \mid x,\pi^*(x)\big)$$

$$\left(r(x,a^{**}) + \int u^*(y) Q''(dy \mid x,a^{**}) \right.$$

$$\left. = r\big(x,\pi^{**}(x)\big) + \int u^*(y) Q''\big(dy \mid x,\pi^{**}(x)\big) \right).$$

We assumed $\pi^\star(x) = \left(\pi_i^\star \left(x_{\widetilde{N}(i)} \right), \ i \in V \right) \in LS$, where $\pi_i^\star \left(x_{\widetilde{N}(i)} \right)$ $\in A_i \left(x_{\widetilde{N}(i)} \right)$. Therefore, because the values of the selector are deterministic, we have proved $\pi^\star(x) \in LS_S$. Similarly, $\pi^{\star\star}(x) \in LS_S$.

The next lemmata can be found in [GS72b].

Lemma 3.58. If A is a compact metric space, $A \colon X \to 2^A - \{\emptyset\}$ is lower semicontinuous, and the function $u(x, a)$ is bounded and continuous on $\bar{\kappa}$, then the function $u^\star(x) := \sup_{a \in A(x)} u(x, a)$ is lower semicontinuous on X.   ⊙

Lemma 3.59. If A is a compact metric space, $A \colon X \to 2^A - \{\emptyset\}$ is upper semicontinuous, and the function $u(x, a)$ is bounded and upper semicontinuous on κ, then the function $u^\star(x) := \sup_{a \in A(x)} u(x, a)$ is upper semicontinuous.   ⊙

Theorem 3.60. Let A be a compact metric space, assume that $A \colon X \to 2^A - \{\emptyset\}$ is continuous, and Assumption 3.55 or 3.56 holds. Assume further that

  1) The function $r(x, a)$, $(x, a) \in \bar{\kappa}$ is continuous;
  2) The transition kernel $Q(\cdot \mid x, a)$, $(x, a) \in \kappa$ is weakly continuous.

Then an optimal strategy $\pi^\star$ exists in LS_S. The function $\pi^\star \colon X \to A$ can be selected as being in Baire class 1.   ⊙

Proof: Our proof of the theorem borrows arguments from Theorem 3 in [GS72b]:

Conditions 1, 2 and Lemmas 3.58, 3.59 imply that the operator $\mathfrak{U}'$ (operator $\mathfrak{U}''$) transforms the Banach space $C(X) \subset M(X)$ of bounded continuous functions on X into itself. We have to show that the map A'_u is upper semicontinuous for every function $u \in C(X)$ (semicontinuity of

A''_u will be proved similarly). For this purpose, it suffices to check according to [Kur69, p. 61] that for every function $u \in C(X)$ the following holds: If $(x^n, a^n) \to (x, a)$ as $n \to \infty$ and $a^n \in A'_u(x^n)$ $(n = 1, 2, \ldots)$, then $a \in A'_u(x)$. (That the sets $A'_u(x)$ $(A''_u(x))$ are nonempty and closed is obvious.) From upper semicontinuity of the map A, it follows that $a \in A(x)$. From $a^n \in A'_u(x^n)$ we have

$$r(x^n, a^n) + \int u(y) Q'(dy \mid x^n, a^n)$$
$$= \max_{a' \in A(x^n)} \left\{ r(x^n, a') + \int u(y) Q'(dy \mid x^n, a') \right\}.$$

Both the left — and the right — hand side of this equality are continuous functions. Passing to the limit for $n \to \infty$ we obtain

$$r(x, a) + \int u(y) Q'(dy \mid x, a) = \max_{a' \in A(x)} \left\{ r(x, a') + \int u(y) Q'(dy \mid x, a') \right\},$$

i.e., $a \in A'_u(x)$.

From Theorem 3.57 an optimal strategy exists and according to the Theorem of Choice for semicontinuous maps, a function $\pi^\star \colon X \to A$ can be selected from Baire class 1. We assumed

$$\pi^\star(x) = \left(\pi_i^\star \left(x_{\widetilde{N}(i)} \right), i \in V \right) \in LS,$$

where $\pi_i^\star \left(x_{\widetilde{N}(i)} \right) \in A_i \left(x_{\widetilde{N}(i)} \right)$, and the measurable selector is a deterministic function.

Therefore, $\pi^\star(x) \in LS_S$. □

3.5.2 Separable reward functions and global optimality of local policies

In this section we consider the control problem for Markov processes with locally interacting synchronous components over $\Gamma = (V, B)$ when the reward function $r(x, a)$ is separable in the sense of the following definition:

Definition 3.61. A reward function $r\colon \bar{\kappa} = \{(x, a)\colon x \in X,\ a \in A(x)\} \subset X \times A \to \mathbb{R}$ is separable (with respect to $\Gamma = (V, B)$) if

$$r(x, a) = \sum_{i \in V} r_i\left(x_{\widetilde{N}(i)}, a_i\right),$$

i.e., if it is a sum of locally defined measurable bounded reward functions $r_i\left(x_{\widetilde{N}(i)}, a_i\right)$. $\odot$

In this section we assume that the control sets are of product structure (see Definition 2.16) but not necessarily require that they are locally determined. Because we assumed time independent decision set it follows:

$$A^t(x) = A(x) = \underset{i \in V}{\times}\, A_i(x) \text{ for all } t, \text{ i.e., } A_i(x) \text{ may depend on the}$$

complete state x and is independent of time. We recall the accordingly defined nonlocal subclasses of general policies.

Definition 3.62. (1) For any vertex $i \in V$ and any system configuration (state) $\xi^t = x$, let $A_i(x)$ be the time invariant set of admissible actions at node i.

(2) A randomized strategy π is called admissible if $\pi_i^t(\cdot \mid x^0, a^0, \dots$
$\dots, x^{t-1}, a^{t-1}, x^t)$ is a probability measure on $\left(A_i\left(x^t\right), \mathfrak{A}_i\left(x^t\right)\right)$ and measurably dependent on the history $h_i^t = \left(x^0, a^0, \dots, x^{t-1}, a^{t-1}, x^t\right)$ for every $t \in \mathbb{N}$.

The class of all admissible strategies is denoted by Π; the subclass of admissible Markov strategies by Π_M.

(3) By Π_D we denote the class of admissible stationary Markov deterministic strategies with time invariant restriction sets: $\pi^t = \pi^{t'}$ for all $t, t' \in \mathbb{N}$ on $A(x) = \underset{i \in V}{\times}\, A_i(x)$. $\odot$

We first state an analogy to Theorem 3.51, the proof of which is similar to the one presented there.

Theorem 3.63. Let there exist a constant g and a bounded Borel function $v(x)$ on X such that

$$g + v(x) = \sup_{a \in A(x)} \left\{ r(x, a) + \int_X v(y) Q(dy \mid x, a) \right\}, \qquad x \in X. \quad (3.39)$$

Then

$$\sup_{\pi \in \Pi} \phi(x, \pi) \leq g, \qquad x \in X.$$

Moreover, if for a strategy $\pi^\star \in \Pi$

$$g + v(x) = r\big(x, \pi^\star(x)\big) + \int_X v(y) Q\big(dy \mid x, \pi^\star(x)\big), \qquad x \in X,$$

then $\phi(x, \pi^\star) \equiv g$, and $\pi^\star$ is optimal. $\qquad \odot$

Corollary 3.64. If (3.39) holds and a strategy $\pi^\star \in LS_D \subseteq \Pi$ fulfills

$$g + v(x) = r\big(x, \pi^\star(x)\big) + \int v(y) Q\big(dy \mid x, \pi^\star(x)\big), \qquad x \in X,$$

then there exist admissible deterministic local policies that are optimal in S.$\odot$

In the present context, Assumption 3.55 takes the following form:

Assumption 3.65. $Q = \prod_{j \in V} Q_j$ is according to Definition 3.10 a local and synchronous Markov kernel for (ξ, π) and the admissible control sets are independent of time.

There exist non-negative measures μ_i on $(X_i, \mathfrak{X}_i)$ with $\mu_i(X_i) > 0$, $\forall\, i \in V$, such that:

$$\mu_i(B_i) \leq Q\Big(B_i \mid x_{\widetilde{N}(i)}, a_i \Big), \qquad B_i \in \mathfrak{X}_i, \; \Big(x_{\widetilde{N}(i)}, a_i \Big) \in \kappa_i, \; i \in V. \qquad \odot$$

Let $S(X) \subseteq M(X)$ denote the metric space of real valued separable functions $v(x) = \sum_{i \in V} v_i(x_i)$ according to Definition 3.61 with norm $\|v\| = \sum_{i \in V} \sup_{x_i \in X_i} |v_i(x_i)|$. Define the operator $\widetilde{\mathfrak{U}}'$ on $S(X)$:

$$\widetilde{\mathfrak{U}}'v(x) = \sup_{a \in A} \left\{ \sum_{i \in V} r_i\left(x_{\widetilde{N}(i)}, a_i\right) + \int_X v(y)Q'(dy \mid x, a) \right\},$$

where

$$Q'(B \mid x, a) = Q(B \mid x, a) - \prod_{i \in V} \mu_i(B_i).$$

Let

$$\widetilde{A}'_v(x) = \left\{ a : a \in A, \ \widetilde{\mathfrak{U}}'v(x) = \sum_{i \in V} r_i\left(x_{\widetilde{N}(i)}, a_i\right) + \int_X v(y)Q'(dy \mid x, a) \right\}$$

and for measurable $v_i \colon X_i \to \mathbb{R}$

$$U_i^{v_i}\left(x_{\widetilde{N}(i)}, a_i\right) = r_i\left(x_{\widetilde{N}(i)}, a_i\right) +$$

$$+ \int_{X_i} v_i(y_i) \left[Q_i\left(dy_i \mid x_{\widetilde{N}(i)}, a_i\right) - \mu_i(dy_i) \prod_{j \colon \left\{ \substack{j \in V; \\ j \neq i.} \right.} \mu_j(X_j) \right].$$

Theorem 3.66. Let the decision set A be a compact metric space with countable basis and assume that $A \colon X \to 2^A$ is continuous. Assume that Assumption 3.65 holds and additionally

1) The functions $r_i\left(x_{\widetilde{N}(i)}, a_i\right)$ are continuous on κ_i.

2) The transition probabilities $Q\left(B_i \mid x_{\widetilde{N}(i)}, a_i\right)$ are weakly continuous on κ_i.

Then an optimal strategy in class LS_D exists. $\qquad \odot$

Proof: The operator $\widetilde{\mathfrak{A}}'$ is monotone increasing:

$$v^1(x) \leq v^2(x), \quad x \in X \quad \Longrightarrow \quad \widetilde{\mathfrak{A}}'v^1(x) \leq \widetilde{\mathfrak{A}}'v^2(x), \quad x \in X;$$

and from

$$\widetilde{\mathfrak{A}}'\big(v(x) + c\big) = \sup_{a \in A(x)} \left\{ \sum_{i \in V} r_i\Big(x_{\widetilde{N}(i)}, a_i\Big) + \int_X v(y)Q'(dy \mid x, a) + \right.$$

$$\left. + c \int_X Q'(dy \mid x, a) \right\}$$

$$= \widetilde{\mathfrak{A}}'v(x) + \alpha c$$

$\widetilde{\mathfrak{A}}'$ fulfills: For any nonnegative constant c is

$$\widetilde{\mathfrak{A}}'\big(v(x) + c\big) = \widetilde{\mathfrak{A}}'v(x) + \alpha c, \qquad x \in X, \quad \alpha = 1 - \prod_{i \in V} \mu_i(X_i) < 1.$$

As $\alpha < 1$, these properties guarantee that $\widetilde{\mathfrak{A}}'$ is a contraction. Indeed, applying the operator $\widetilde{\mathfrak{A}}'$ on both sides of the inequality

$$v^1(x) \leq v^2(x) + \rho\big(v^1, v^2\big)$$

we have

$$\widetilde{\mathfrak{A}}'v^1(x) \leq \widetilde{\mathfrak{A}}'\Big(v^2(x) + \rho\big(v^1, v^2\big)\Big)$$

$$= \widetilde{\mathfrak{A}}'v^2(x) + \alpha\rho\big(v^1, v^2\big).$$

Therefore

$$\widetilde{\mathfrak{A}}'v^1(x) - \widetilde{\mathfrak{A}}'v^2(x) \leq \alpha\rho\big(v^1, v^2\big).$$

Interchanging v^1 and v^2, we have

$$\big|\widetilde{\mathfrak{A}}'v^1(x) - \widetilde{\mathfrak{A}}'v^2(x)\big| \leq \alpha\rho\big(v^1, v^2\big), \qquad x \in X,$$

i.e.,

$$\rho\big(\widetilde{\mathfrak{A}}'v^1, \widetilde{\mathfrak{A}}'v^2\big) \leq \alpha\rho\big(v^1, v^2\big),$$

and $\widetilde{\mathfrak{U}}'$ is a contraction. Banach's fixed point theorem implies that in $S(X)$ exists a function $v^{\star}(x)$ such that

$$v^{\star}(x) = \sup_{a \in A(x)} \left\{ \sum_{i \in V} r_i\left(x_{\widetilde{N}(i)}, a_i\right) + \int_X v^{\star}(y) \mathsf{Q}'(dy \mid x, a) \right\}, \qquad x \in X.$$

The definition of $\widetilde{\mathfrak{U}}'$ and the separability of $r(\cdot, \cdot)$ yield

$$\widetilde{\mathfrak{U}}' v^{\star}(x) = \sum_{i \in V} \sup_{a_i \in A_i(x)} U_i^{v_i^{\star}}\left(x_{\widetilde{N}(i)}, a_i\right).$$

From condition 1) and 2) of the theorem it follows that there exist measurable functions $a_i^{\star} = a_i^{\star}\left(x_{\widetilde{N}(i)}\right)$ such that

$$\sup_{a_i \in A_i(x)} U_i^{v_i^{\star}}\left(x_{\widetilde{N}(i)}, a_i\right) = U_i^{v_i^{\star}}\left(x_{\widetilde{N}(i)}, a_i^{\star}\left(x_{\widetilde{N}(i)}\right)\right).$$

Therefore there exists a function $\pi^{\star} = \left\{ \pi_i^{\star}\left(x_{\widetilde{N}(i)}\right) \right\} \in LS_S$ such that

$$\widetilde{\mathfrak{U}}' v^{\star}(x) = \sum_{i \in V} r_i\left(x_{\widetilde{N}(i)}, \pi_i^{\star}\left(x_{\widetilde{N}(i)}\right)\right) + \int_X v^{\star}(y) \mathsf{Q}'\left(dy \mid x, \pi^{\star}(x)\right)$$

and

$$v^{\star}(x) = \sum_{i \in V} r_i\left(x_{\widetilde{N}(i)}, \pi_i^{\star}\left(x_{\widetilde{N}(i)}\right)\right) + \int_X v^{\star}(y) \mathsf{Q}'\left(dy \mid x, \pi^{\star}(x)\right)$$

or

$$g + v^{\star}(x) = \sum_{i \in V} r_i\left(x_{\widetilde{N}(i)}, \pi_i^{\star}\left(x_{\widetilde{N}(i)}\right)\right) + \int_X v^{\star}(y) \mathsf{Q}\left(dy \mid x, \pi^{\star}(x)\right),$$

where

$$g = \int_X v^{\star}(y) \prod_{i \in V} \mu_i(dy_i)$$

$$= \sum_{i \in V} \int_{X_i} v_i^{\star}(y_i) \mu_i(dy_i) \prod_{j:\ \left\{\substack{j \in V; \\ j \neq i.}\right.} \mu_j(X_j).$$

Therefore, the conditions of Theorem 3.63 are satisfied and strategy $\pi^\star \in LS_S$ is optimal. □

Chapter 4

SEQUENTIAL STOCHASTIC GAMES WITH DISTRIBUTED PLAYERS ON GRAPHS

In this chapter we develop the concept of locally acting distributed players in sequential stochastic games with general compact state and action spaces extending [CDK04]). The state transition function for the system is Markov in time and of local structure, which results in Markov properties in space and time for the describing processes. We prove that we can reduce optimality problems for local strategies to only considering Markov strategies. We further prove the existence of optimal strategies and the existence of a value for the game with respect to the asymptotic average reward criterion.

Stochastic games will be understood in the following as multiperson stochastic sequential games with an infinite horizon. The description of the behavior of such systems is therefore by construction of suitable stochastic processes with discrete time scale.

There is a wide area of applications for the games we have in mind. Many phenomena of management science and economics, biology and psychology, and especially military affairs are modeled by using stochastic sequential games with very different frameworks. A short introduction with many specific examples can be found in [HS84, Chapter 9].

There is recent new interest in many-player stochastic games as an essential ingredient of the control procedures for large networks of interconnected stations, especially in network nodes of computer and communications networks that share common resources. The most prominent example is the Internet. Classical control of queueing systems is modeled and performed either by using stochastic dynamic optimization or by using two-person games, where one of the players represents the nature that perturbs the functioning of the system: An introduction into the field is [FV96], which summarizes those techniques as *Competitive Markov Decision Processes*. For applications of game theoretical methods to queueing systems and small networks, see [Alt96, Alt99, AH95], and the references therein.

However, in case of the large networks that we have in mind, typically a large number of players (nodes of the networks) interact and compete for the network's resources. These players can act cooperatively or noncooperatively. In the first case this leads to building coalitions, which usually have partly antagonistic criteria for the performance measures of the network. The introduction to multiperson games in [HS84, Chapter 9] is centered around noncooperative multiperson games. General schemes for cooperation under incomplete information with bargaining are described from an abstract point of view in [Mye84]. A monography on classical topics of stochastic games is [MS96]. A more recent survey including problems of multiperson games is [NS99]. A classical paper on the general subject with strong connections to stochastic dynamic optimization is [Fed78]. In [RF91], algorithms for stochastic games with finite state and action spaces and finite time horizon are described.

More recent work on stochastic network control as a multiperson game are [KL95] and [Yao95]. In [KL95], the situation of players in a network in equilibrium is investigated and it is proved that individual conditional best reply strategies given the strategies of the other customers yield a global Nash equilibrium for the network. Due to the equilibrium assumption, this problem can be solved by means of static game theory. The noteworthy observation is: There is no centralized decision maker who controls the network's resources, and/or admits customers to the network on the basis of a global admission regime. In [Yao95], a certain dynamical viewpoint is introduced in that algorithms are described that can be applied by the players over successive time steps in a way that the joint strategy of the players approach an equilibrium strategy. The network is then considered to be in equilibrium.

The standard models of multiperson systems that compete and/or cooperate for common resources allow the players to make their decisions on the basis of either the whole history of the system, or (as can be proved to be almost similarly effective) on the basis of the knowledge of the complete actual state of the system only. However, for large networks this seems to be an optimistic or unrealistic assumption because any player in real large networks will be able to take into consideration for his decision making only what is known about the state of his (suitably defined) neighborhood. It follows that in a natural way there emerges a neighborhood structure in the set of decision makers (players), and therefore in the set of nodes, where the players reside. Such neighborhoods are usually determined by a neighborhood graph, which in our case will be undirected, but this restriction can be removed easily.

Models from economics where neighborhood systems emerge and are developed as structural property to determine equilibrium behavior of the system and to construct efficient control mechanisms for optimizing the system are described in Example 3.11. Neighborhood systems are well known structures in stochastic process theory and their applications: They determine which coordinates interact in the interacting processes. We are faced here with similar problems with locality and synchronization, when trying to control the space-time behavior of interacting players in networks in a game theoretical interaction.

With respect to control of large networks, it is pointed out in [Alt96] and [KL95] that there is a strong need and "growing interest in situations of several controllers and several objectives, which gives rise to noncooperative models". Introducing coalitions into the game structure here can therefore be seen as continuing the study of [Alt96] (and similar papers) where the interplay of, say, an admission controller to a queue, and a service controller of that queue is considered. Here we would interpret one coalition as the set of admission controllers to a data transmission and communication network and the other to be the set of service controllers in that network. While usually the first coalition is interested in small transmission delay, the second coalition has as main objective high throughputs.

Note that for such control problems, it is again natural to assume that the controllers use only information available in their neighborhood. In high-speed networks it is almost necessary to restrict oneself to use only policies depending on neighborhood states, because too much information incorporated would diminish the performance of the control regime. This is applied, e.g., in the definition of the *Dynamic Alternative Routing* algorithm (see [GKK95]), and further developed in the suggested *Balanced Dynamic Alternative Routing* algorithm, see [AKU02].

Starting from these observations and similarly from other network structures with locality properties where interacting players occur, our aim in this chapter is to develop a general model for describing the interaction of players and coalitions that live on graph structured networks and have to optimize their behavior in sequential stochastic games. This leads immediately to the main questions of the existence of optimal strategies and a value of the games, and to the question of whether these optimal strategies can be chosen from easy to handle subclasses of strategies. The main point is that we shall consider only local strategies as being suitable for the situations described above. Within this class we shall prove the required existence theorems and structural properties, allowing for the technical framework rather general assumptions.

Summarizing: In this chapter we introduce the concept of locality into general stochastic games. This can also be quoted as another concept of incomplete information, different from that usually dealt with; see, e.g., [Mye84], where each player has complete knowledge about his own *type* (which summarizes his properties of relevance to the development of the system), and has only a (probabilistic) guess about the *type* of the other players.

4.1   DISTRIBUTED PLAYERS THAT ACT LOCALLY

For systems with locally interacting coordinates, the interaction structure is defined via an undirected finite neighborhood graph. For the relevant definitions and general notations, recall Section 3.1 for product spaces and Section 3.2 for the fundamentals of interacting processes. We always assume that the processes that occur are locally structured and synchronized with respect to transitions of the coordinates (Definition 3.10) and the sequential decision making.

4.1.1   Distributed players and local policies for coalitions

We next describe $|V|$-person (players) stochastic games. The players are located at the vertices of the interaction graph Γ; the edges represent the connections between the players. Our aim is to model the behavior of two fixed coalitions of the players in a way that the stochastic process model for the interaction of players resembles Markovian neighborhood systems, and that the development of the game over time is a Markov process with a transition mechanism similar to that described in Definition 3.10.

This will result in the notion of a controlled Markov process with locally interacting synchronous components, shortly: A controlled time dependent Markov random field. For easier understanding of the modeling procedure,, we can think that some time dependent Markov random field $\xi = \{\xi^t : t \in \mathbb{N}\}$ according to Definition 3.10 is given as our starting point, which describes the behavior of the $|V|$ persons that interact locally and synchronously but do not have the possibility of making decisions. ξ will be equipped in the sequel with a control structure that governs the transition kernel of ξ. As usual in multiperson games, the control and decision mechanism is synchronized and will be localized according to the definitions below, which resemble (3.3) and therefore will result in a Markov random field behavior in space for a fixed time instant. Over time we construct as usual a (Markovian) interaction process.

Our players are located at the edges of the finite graph $\Gamma = (V, B)$, so the coalitions that the players build can be described by a decomposition of the node set $V = V_1 \cup V_2$, $V_1, V_2 \neq \emptyset$, $V_1 \cap V_2 = \emptyset$. V_1 is the set of vertices where players of coalition K_1 reside; V_2 is a set of vertices for players of coalition K_2.

For a shorthand notation, we shall often identify a vertex and the player who resides there. For the general notation of control actions, recall Definition 3.16.

Similar to Definition 3.17 we denote by α_i^t the action chosen by the player (decision maker) at node i at time t, $\alpha^t := (\alpha_i^t : i \in V)$ the joint decision vector of all players at time $t \in N$.

Definition 4.1. (1) A randomized strategy (or policy) π for coalition K_1 with interacting components is defined as a vector of coordinate policies $\pi = (\pi_i, \ i \in V_1)$, where for node i $\pi_i = \{\pi_i^0, \ldots, \pi_i^t, \ldots\}$ is a sequence of transition probabilities $\pi_i^t = \pi_i^t(\cdot \mid x^0, a^0, \ldots, x^{t-1}, a^{t-1}, x^t)$.

Similarly a strategy (or policy) γ for coalition K_2 is $\gamma = (\gamma_j, \ j \in V_2)$ where for node j $\gamma_j = \{\gamma_j^0, \ldots, \gamma_j^t, \ldots\}$ is a sequence of transition probabilities $\gamma_j^t = \gamma_j^t(\cdot \mid x^0, a^0, \ldots, x^{t-1}, a^{t-1}, x^t)$.

So for any history $h^t = (x^0, a^0, \ldots, x^{t-1}, a^{t-1}, x^t)$ of the system up to time t π_i^t is a probability measure on $(A_i, \mathfrak{A}_i)$ and γ_j^t is a probability measure on $(A_j, \mathfrak{A}_j)$, which are measurably dependent on the history $h^t = (x^0, a^0, \ldots, x^{t-1}, a^{t-1}, x^t)$ up to time t.

We therefore have

$$\Pr\left\{\alpha_i^t \in B_i \mid \xi^0 = x^0, \alpha^0 = a^0, \ldots, \xi^{t-1} = x^{t-1}, \alpha^{t-1} = a^{t-1}, \xi^t = x^t\right\}$$
$$= \pi_i^t\left(B_i \mid x^0, a^0, \ldots, x^{t-1}, a^{t-1}, x^t\right) \quad \forall \ B_i \in \mathfrak{A}_i, \ i \in V_1,$$
$$\Pr\left\{\alpha_j^t \in B_j \mid \xi^0 = x^0, \alpha^0 = a^0, \ldots, \xi^{t-1} = x^{t-1}, \alpha^{t-1} = a^{t-1}, \xi^t = x^t\right\}$$
$$= \gamma_j^t\left(B_j \mid x^0, a^0, \ldots, x^{t-1}, a^{t-1}, x^t\right) \quad \forall \ B_j \in \mathfrak{A}_j, \ j \in V_2.$$

(2) We assume that the decisions of the players and the coalitions are synchronized according to the following scheme.

$$\Pr\left\{\alpha^t \in \underset{i\in V}{\times} B_i \mid \xi^0 = x^0, \alpha^0 = a^0, \dots \right.$$

$$\left. \dots, \xi^{t-1} = x^{t-1}, \alpha^{t-1} = a^{t-1}, \xi^t = x^t\right\}$$

$$= \prod_{i\in V} \Pr\left\{\alpha_i^t \in B_i \mid \xi^0 = x^0, \alpha^0 = a^0, \dots \right.$$

$$\left. \dots, \xi^{t-1} = x^{t-1}, \alpha^{t-1} = a^{t-1}, \xi^t = x^t\right\}$$

$$= \prod_{i\in V_1} \pi_i^t\left(B_i \mid x^0, a^0, \dots, x^{t-1}, a^{t-1}, x^t\right) \times$$

$$\times \prod_{j\in V_2} \gamma_j^t\left(B_j \mid x^0, a^0, \dots, x^{t-1}, a^{t-1}, x^t\right),$$

$$B_k \in \mathfrak{A}_k, \quad a^s \in A, \quad x^s \in X. \quad \odot$$

The definition above prescribes according to the saying in (3.4) and (3.8) a synchronous control kernel. Note however, that this form of synchronization is a standard assumption for decision making in discrete time sequential games. Even games with perfect information may be considered as allowing at each decision instant for one of the customers only one "decision"; see [Fed78, Section 5], which formally results in the general synchronization.

We now introduce various forms of structured strategies. We shall give the definition only for coalition K_1. Strategies for coalition K_2 are defined similarly. We first fix the following assumption for this subsection:

Assumption 4.2. The admissible control sets are independent of the time of decision:

$$A(x) := A^t(x) = \underset{i\in V}{\times} A_i^t\left(x_{\widetilde{N}(i)}\right) =: \underset{i\in V}{\times} A_i\left(x_{\widetilde{N}(i)}\right), \quad \forall\, t \in \mathbb{N}. \quad \odot$$

Definition 4.3. (1) For any player (vertex) i an admissible local strategy $\pi_i = \{\pi_i^t, t \in \mathbb{N}\}$ is defined according to Definition 3.18 **(1)**. $\pi = (\pi_i : i \in V_1)$ is said to be locally admissible for coalition K_1 if all π_i are locally admissible, $i \in V_1$.

(2) An admissible local strategy $\pi = (\pi_i, \ i \in V_1)$ is called admissible local Markov strategy for coalition K_1 if

$$\pi_i^t\left(\cdot \mid x_{\widetilde{N}(i)}^0, a_i^0, \ldots, x_{\widetilde{N}(i)}^{t-1}, a_i^{t-1}, x_{\widetilde{N}(i)}^t\right) = \pi_i^t\left(\cdot \mid x_{\widetilde{N}(i)}^t\right), \quad i \in V_1.$$

(3) An admissible local Markov strategy $\pi = (\pi_i, \ i \in V_1)$ is called admissible stationary local strategy for coalition K_1 if $\pi_i^{t'}\left(\cdot \mid x_{\widetilde{N}(i)}\right) = \pi_i^{t''}\left(\cdot \mid x_{\widetilde{N}(i)}\right)$, $i \in V_1$, for all t' and t'' and all x.

(4) An admissible stationary local strategy $\pi = (\pi_i, \ i \in V_1)$ is called admissible deterministic (nonrandomized) stationary local strategy for K_1 if $\pi_i^t\left(\cdot \mid x_{\widetilde{N}(i)}\right)$, $i \in V$, are one-point measures on $A_i^t\left(x_{\widetilde{N}(i)}\right)$, $i \in V$, for each $x \in X$.

The class of all admissible local strategies is denoted by LS; the subclass of admissible local Markov strategies by LS_M. By $LS_S \subseteq LS_M$, we denote the class of admissible stationary local strategies, and by $LS_D \subseteq LS_S$, the set of admissible deterministic stationary local strategies.

If it is clear from the context, we shall henceforth omit the *"admissible"* in the description of the strategies under consideration. $\odot$

Similar to (3.3) and (3.8) we construct our process in a way that the Markov property holds: The law of motion of the system is characterized by a set of time invariant transition probabilities. Whenever the state of the system is $\xi^t = x^t$ and decision $\alpha^t = a^t$ is taken, then the transition probability is $\Pr\left\{\xi^{t+1} \in C \mid \xi^t = x^t, \alpha^t = a^t\right\} =: \mathsf{Q}(C \mid x^t, a^t)$, and therefore independent of the past given the present (generalized) state. We further assume that this transition probability is independent of t, and that the motion is homogeneous in time.

Applying a game theoretic control policy from two coalitions to a time dependent Markov random field from Definition 3.10, we shall call the triple (ξ, π, γ) a controlled version of ξ using strategies π and γ. The controlled process $\xi = (\xi^t)$ in general is not Markovian because at first the sequence (α^t) of decisions depends not only on the actual local states $x^t_{\widetilde{N}(i)}$, $i \in V$, but on the previous local states $x^0_{\widetilde{N}(i)}, \ldots, x^{t-1}_{\widetilde{N}(i)}$ as well.

An immediate consequence of the Definition 4.3 will be that if Markov strategies π and γ are applied to control the time dependent Markov random field, then we obtain a Markov process ξ from (ξ, π, γ).

Definition 4.4. A triple (ξ, π, γ) is called a controlled stochastic process with locally interacting synchronous components with respect to the finite interaction graph $\Gamma = (V, B)$ with coalitions K_1 (located at V_1) and K_2 (located at V_2), $V = V_1 \cup V_2$, $V_1 \cap V_2 = \emptyset$, if $\xi = (\xi^t\colon t \in \mathbb{N})$ is a stochastic process with state space $X = \underset{i \in V}{\times} X_i$, $\pi = (\pi_i\colon i \in V_1)$ and $\gamma = (\gamma_j\colon j \in V_2)$ are admissible local strategies, and the transitions of ξ are determined as given below in (4.1). If the strategies (π, γ) of both coalitions are stationary local Markov, then the triple (ξ, π, γ) is called a controlled Markov process with locally interacting synchronous components and has a time invariant transition law.

The rationale behind the construction of this transition law is along the following lines.

$$\Pr\left\{\xi^{t+1}_K \in C_K \mid \xi^0 = x^0, \alpha^0 = a^0, \ldots\right.$$
$$\left.\ldots, \xi^{t-1} = x^{t-1}, \alpha^{t-1} = a^{t-1}, \xi^t = y, \alpha^t = a\right\}$$
$$= \Pr\left\{\xi^{t+1}_K \in C_K \mid \xi^0 = x^0, \ldots, \xi^{t-1} = x^{t-1}, \xi^t = y, \alpha^t = a\right\}$$
$$= \Pr\left\{\xi^{t+1}_K \in C_K \mid \xi^t = y, \alpha^t = a\right\}$$
$$= \prod_{j \in K} \Pr\left\{\xi^{t+1}_j \in C_j \mid \xi^t = y, \alpha^t = a\right\}$$
$$= \prod_{j \in K} \Pr\left\{\xi^{t+1}_j \in C_j \mid \xi^t_{\widetilde{N}(j)} = y_{\widetilde{N}(j)}, \alpha^t_j = a_j\right\}$$

$$= \prod_{j \in K} Q_j \Big(C_j \mid y_{\widetilde{N}(j)}, a_j \Big)$$

$$= Q_K(C_K \mid y, a), \qquad K \subseteq V, \quad y \in X, \quad a_j \in A_j \Big(y_{\widetilde{N}(j)} \Big). \quad (4.1)$$

If $K = V$, we shall write $Q_V(C_V \mid y, a) =: Q(C \mid y, a)$.

The Markov kernel $Q = \prod_{j \in V} Q_j$ is said to be local and synchronous. ξ will then shortly be called a controlled time dependent random field. ⊙

Some remarks on the modeling principle behind this definition are given after Definition 3.21 for the case of standard interacting processes and apply here as well.

At every time instant $t \in \mathbb{N}$ the players from both coalitions make decisions on the basis of the complete information about the state history of their full neighborhoods and of all their own preceding decisions. It is possible that only players of one's own coalition or players of both coalitions belong to a player's neighborhood. After the joint decision of all players, the i-th player (at the i-th vertex) receives from the j-th player a (positive or negative) payment (reward) $r_{ij}(x_i, x_j, a_i, a_j)$, $x_i \in X_i$, $x_j \in X_j$, $a_i \in A_i$, $a_j \in A_j$, $i \in V$, $j \in N(i)$, where we assume $r_{ij} = -r_{ji}$. Thereafter the state of the system changes according to the transition law Q in (4.1).

The overall profits of coalitions K_1 and K_2 at each time instant are $r_1(x, a) = \sum_{i \in V_1} \sum_{j \in N(i)} r_{ij}(x_i, x_j, a_i, a_j)$ and $r_2(x, a) = \sum_{j \in V_2} \sum_{i \in N(j)} r_{ji}(x_j, x_i, a_j, a_i)$. We have a zero-sum game: $r_1(x, a) = -r_2(x, a)$, and the sum of payments inside each coalition equals zero. Therefore, we write $r(x, a) = r_1(x, a) = -r_2(x, a)$.

We assume that the reward function r and the transition probabilities Q are Borel measurable, and that furthermore

$$\big| r(x, a) \big| \leq L < \infty, \qquad x \in X, \quad a \in A.$$

The tuple (X, A, Q, V_1, V_2, r) defines the stochastic game on neighborhood graph Γ.

The asymptotic average mean reward for initial state $\xi_V^0 = x^0$ of ξ, and when policies π, γ are applied is

$$\phi(x^0, \pi, \gamma) = \liminf_{n \to \infty} \frac{1}{n+1} \mathsf{E}_{\pi,\gamma}^{x^0} \sum_{k=0}^{n} r(\xi^k, \alpha^k),$$

where $\mathsf{E}_{\pi,\gamma}^{x^0}$ is the expectation associated with the controlled process (ξ, π, γ) under $\xi^0 = x^0$.

Definition 4.5. The strategy $\pi^\star \in LS$ is called optimal in the class LS of admissible local strategies for coalition K_1 if for all admissible local strategies γ' of coalition K_2 and all initial states $x \in X$ of the system

$$\phi(x, \pi^\star, \gamma') \geq \inf_{\gamma \in LS} \sup_{\pi \in LS} \phi(x, \pi, \gamma).$$

The strategy $\gamma^\star \in LS$ is called optimal in the class LS of admissible local strategies for coalition K_2 if for all admissible local strategies π' of coalition K_1 and all initial states $x \in X$ of the system

$$\phi(x, \pi', \gamma^\star) \leq \sup_{\pi \in LS} \inf_{\gamma \in LS} \phi(x, \pi, \gamma).$$

The stochastic game has a value if

$$\inf_{\gamma \in LS} \sup_{\pi \in LS} \phi(x, \pi, \gamma) = \sup_{\pi \in LS} \inf_{\gamma \in LS} \phi(x, \pi, \gamma), \qquad x \in X. \qquad \odot$$

4.1.2 Relation to stochastic dynamic optimization: Reduction to Markov strategies

To study a stochastic game on a graph, we will introduce a reduced model (R-model) similarly to the case of stochastic two-person games [Sha53, MP70, Gub72]. The idea behind this procedure is in the two-person case, to fix the strategy of one of the players, say, player 2, and incorporate the then known stochastic effects of his decision into the transition kernel for the state transition mechanism. Then theorems, methods, and algorithms of standard stochastic dynamic optimization can be applied to optimize the decisions of player 1 to find at least a conditioned optimal behavior.

In the situation of our n-person games with two coalitions we mimic this by fixing the strategies for one coalition and then will end in the framework of control and optimization of Markov interacting processes, or time dependent random fields.

If A is a compact metric space, we denote by P_A the space of all probability measures on the Borel sets of A. It is well known that P_A, endowed with the weak topology, is a compact metric space. Throughout the chapter, the space of probability measures on the Borel subsets of a compact metric space is endowed with the weak topology.

Let $A_{V_i} = \underset{k \in V_i}{\times} A_k$, $i = 1, 2$, and recall $a = (a_{V_1}, a_{V_2})$ and put

$$r(x, \zeta, \lambda) = \iint r(x, a) \, d\zeta(a_{V_1}) \, d\lambda(a_{V_2}),$$

$$Q(\cdot \mid x, \zeta, \lambda) = \iint Q(\cdot \mid x, a) \, d\zeta(a_{V_1}) \, d\lambda(a_{V_2}),$$

$$x \in X, \quad \zeta \in \mathsf{P}_{A_{V_1}}, \quad \lambda \in \mathsf{P}_{A_{V_2}}.$$

Let g be the fixed local stationary Markov strategy of coalition K_2, i.e., $g \colon \widetilde{X}_{V_2} \to \mathsf{P}_{A_{V_2}}$, where $\widetilde{X}_S = \underset{i \in S}{\times} X_{\widetilde{N}(i)}$.

The R-model can be looked upon as a controlled Markov random field with discrete time scale and with compact phase space X and with compact action space A_{V_1} as defined in Section 3.5. The reward function r' and the transition probability Q' have to be defined as follows:

$$r'(x, \zeta) = r\Big(x, \zeta, g\big(\tilde{x}_{V_2}\big)\Big),$$

$$Q'(C \mid x, \zeta) = Q\Big(C \mid x, \zeta, g\big(\tilde{x}_{V_2}\big)\Big),$$

$$C \in \mathfrak{X}, \quad x \in X, \quad \tilde{x}_{V_2} \in \widetilde{X}_{V_2}, \quad \zeta \in \mathsf{P}_{A_{V_1}},$$

and the actions on nodes of K_2 are formally *"Do nothing!"* If with this assumption π' is a strategy (similar to Definition 4.3 with $V_1 \equiv V$, or according to the definitions from Section 3.5 in the R-model), the asymptotic average mean reward for initial state $\xi_V^0 = x$ of ξ, and when policy π' is applied, is

$$\phi'(x, \pi') = \liminf_{n \to \infty} \frac{1}{n+1} \, \mathsf{E}_{\pi'}^x \sum_{k=0}^{n} r'(\xi^k, \alpha_{V_1}^k),$$

where $\mathsf{E}_{\pi'}^x$ is the expectation associated with the controlled process (ξ, π'); see (2.7).

The following lemma follows immediately from the definition of the R-model, where a decision selects a point in the space of probability measures on A_{V_1}.

Lemma 4.6. Assume a local stationary Markov strategy g of coalition K_2 is fixed. If π is a stationary local Markov strategy of coalition K_1 in the stochastic game then π is a deterministic stationary local strategy in the R-model and $\phi'(x, \pi) = \phi(x, \pi, g)$, $x \in X$.

Conversely, if π is a deterministic stationary local strategy in the R-model then π is a stationary local Markov strategy of coalition K_1 in the stochastic game and $\phi(x, \pi, g) = \phi'(x, \pi)$, $x \in X$. $\odot$

4.2 OPTIMAL STRATEGIES FOR LOCALLY ACTING PLAYERS

4.2.1 Stationary local Markov strategies

We are now ready to prove that at least for any coalition and for any local strategy of this coalition there exists (conditioned on the other coalition playing a fixed Markov strategy) a Markovian local policy that reproduces the long time average reward for the game.

Lemma 4.7. Assume that coalition K_2 applies the fixed local stationary Markov strategy g. If π is any admissible local strategy of coalition K_1 in the stochastic game, then there exists a local Markov strategy $\pi^\star$ of coalition K_1 such that $\phi(x, \pi^\star, g) = \phi(x, \pi, g)$, $x \in X$. $\qquad\qquad\odot$

Proof: The proof will use arguments borrowed from the proofs in the framework of standard controlled processes (see, e.g., Theorem 4.1 in [Str66]) and standard stochastic games (see, e.g., Lemma 3.2 in [MP70]).

Let $e_\pi(\cdot \mid x^0)$ be the conditional distribution of the entire history $h = (x^0, a^0, x^1, a^1, \dots)$ of the system, given the initial state $x^0 \in X$, where coalition K_1 uses π and coalition K_2 uses g. For each $n \geq 0$ the conditional distribution under $e_\pi(\cdot \mid x^0)$ of $a_i^n \in A_i$, given x^n, is denoted by $\overset{\star}{\pi}{}_i^n\left(a_i^n \mid x_{\tilde{N}(i)}^n\right)$, $i \in V_1$, and

$$\overset{\star}{\pi}{}^n\left(a^n \mid \tilde{x}_{V_1}^n\right) = \prod_{i \in V_1} \overset{\star}{\pi}{}_i^n\left(a_i^n \mid x_{\tilde{N}(i)}^n\right).$$

Let $\pi^\star = \left(\overset{\star}{\pi}{}^0, \overset{\star}{\pi}{}^1, \dots\right)$. Then $\pi^\star$ is the required Markov strategy for coalition K_1. To see this, denote the conditional distribution of the entire history $h = (x^0, a^0, x^1, a^1, \dots)$ of the system by $e_{\pi^\star}(\cdot \mid x^0)$, given the initial state is x^0, when coalition K_1 uses $\pi^\star$ and coalition K_2 uses g. First, we will prove that for every bounded measurable function p on $X \times A$, and every $n \geq 0$

$$\int p(x^n, a^n)\, de_\pi(a^n \mid x^0) = \int p(x^n, a^n)\, de_{\pi^\star}(a^n \mid x^0) \qquad (4.2)$$

for every $x^0 \in X$.

The proof proceeds by induction. For $n = 0$, note that $\pi^0 = \overset{\star}{\pi}{}^0$. Consequently,

$$\int p(x^0, a^0)\, de_\pi(a^0 \mid x^0) = \iint p(x^0, a^0)\, d\pi^0\left(a_{V_1}^0 \mid \tilde{x}_{V_1}^0\right) dg\left(\tilde{x}_{V_2}^0\right)\left(a_{V_2}^0\right)$$

$$= \iint p(x^0, a^0)\, d\overset{\star}{\pi}{}^0\left(a_{V_1}^0 \mid \tilde{x}_{V_1}^0\right) dg\left(\tilde{x}_{V_2}^0\right)\left(a_{V_2}^0\right)$$

$$= \int p(x^0, a^0) \, de_{\pi^\star}(a^0 \mid x^0).$$

This proves that (4.2) holds for $n = 0$.

Suppose now that (4.2) holds for all $n < N$ and for all bounded Borel functions on $X \times A$. Below, all conditional expectations are relative to $e_\pi(\cdot \mid x^0)$. We have

$$\int p(x^N, a^N) \, de_\pi(a^n \mid x^0) = \mathsf{E}_{x_0} \, p(\xi^N, \alpha^N)$$

$$= \mathsf{E}_{x_0} \left(\mathsf{E}_{x_0} \left[p(\xi^N, \alpha^N) \mid \xi^N \right] \right)$$

$$= \mathsf{E}_{x_0} \, u(\xi^N),$$

where

$$u(\xi^N) = \mathsf{E}_{x_0} \left[p(\xi^N, \alpha^N) \mid \xi^N \right]$$

$$= \iint p(x^N, a^N) \, d\tilde{\pi}^N \big(a_{V_1}^N \mid \tilde{x}_{V_1}^N\big) \, dg\big(\tilde{x}_{V_2}^N\big) \big(a_{V_2}^N\big).$$

Thus,

$$\int p(x^N, a^N) \, de_\pi(\cdot \mid x^0) = \mathsf{E}_{x_0} \left(\mathsf{E}_{x_0} \left[u(\xi^N) \mid \xi^{N-1}, \alpha^{N-1} \right] \right)$$

$$= \mathsf{E}_{x_0} \, w(\xi^{N-1}, \alpha^{N-1}),$$

where

$$w(\xi^{N-1}, \alpha^{N-1})$$

$$= \mathsf{E}_{x_0} \left[u(\xi^N) \mid \xi^{N-1}, \alpha^{N-1} \right]$$

$$= \iiint p(x^N, a^N) \, d\tilde{\pi}^N \big(a_{V_1}^N \mid \tilde{x}_{V_1}^N\big) \, dg\big(\tilde{x}_{V_2}^N\big) \big(a_{V_2}^N\big) \, d\mathsf{Q}(x^{N-1}, a^{N-1}).$$

Now, the induction hypothesis implies that

$$\mathsf{E}_{x_0} \, w(\xi^{N-1}, \alpha^{N-1}) = \int w(x^{N-1}, a^{N-1}) \, de_\pi(\alpha^{N-1} \mid x^0)$$

$$= \int w(x^{N-1}, a^{N-1}) \, de_{\pi^\star}(\alpha^{N-1} \mid x^0).$$

Consequently,

$$\int p(x^N, a^N) \, de_\pi(a^N \mid x^0)$$

$$= \int w(x^{N-1}, a^{N-1}) \, de_{\pi^\star}(a^N \mid x^0)$$

$$= \iiiint p(x^N, a^N) \, d\overset{\star}{\pi}^N(a_{V_1}^N \mid \tilde{x}_{V_1}^N) \times$$

$$\times \, dg(\tilde{x}_{V_2}^N)(a_{V_2}^N) \, dQ(x^N \mid x^{N-1}, a^{N-1}) \, de_{\pi^\star}(a^{N-1} \mid x^0)$$

$$= \int p(x^N, a^N) \, de_{\pi^\star}(a^N \mid x^0).$$

Thus, (4.2) has been verified for $n = N$. By (4.2) we have

$$\phi(x, \pi, g) = \liminf_{n \to \infty} \frac{1}{n+1} \, \mathsf{E}_{\pi,g} \sum_{k=0}^{n} r(\xi^k, \alpha^k)$$

$$= \liminf_{n \to \infty} \frac{1}{n+1} \, \mathsf{E}_{\pi^\star,g} \sum_{k=0}^{n} r(\xi^k, \alpha^k)$$

$$= \phi(x, \pi^\star, g), \qquad x \in X.$$

The proof is completed.   □

Remark 4.8. A similar statement as for coalition K_1 in Lemma 4.7 holds for policies of coalition K_2.   ⊙

The next two theorems will, under some technical conditions, provide us with the existence of values for games and with optimal stationary local Markov strategies. These conditions will enable us to later prove general existence theorems under mild conditions on the data of the systems.

Theorem 4.9. Assume that for the stochastic game there exist admissible local stationary (Markov) strategies f and g for coalitions K_1 and K_2 respectively, a constant c and a bounded measurable function $v(x)$ on X such that

$$c + v(x)$$
$$= \sup_{\zeta \in P_{A_{V_1}}} \left\{ r\Big(x, \zeta, g(\tilde{x}_{V_2})\Big) + \int v(y) \mathsf{Q}\Big(dy \mid x, \zeta, g(\tilde{x}_{V_2})\Big) \right\}, \quad (4.3)$$

$$c + v(x)$$
$$= \inf_{\delta \in P_{A_{V_2}}} \left\{ r\Big(x, f(\tilde{x}_{V_1}), \delta\Big) + \int v(y) \mathsf{Q}\Big(dy \mid x, f(\tilde{x}_{V_1}), \delta\Big) \right\}, \quad (4.4)$$

for all $x \in X$. Then the stochastic game has a value that is identically equal to c and strategies f and g are optimal in LS for coalitions K_1 and K_2, respectively. $\odot$

Proof: Because we restrict our computations to the set of local admissible strategies LS, so any $\sup_\gamma$, $\inf_\pi$, etc., is to be read with this restriction. Let coalition K_2 be allowed to use only strategy g. Then the optimality equation (3.36) in Theorem 3.53 implies that for any strategy π' in the R-model $\phi'(x, \pi') \leq c$ holds, for all $x \in X$. From this and using Lemma 4.7, we then obtain for any strategy π of coalition K_1 $\phi(x, \pi, g) \leq c$, $x \in X$, in the stochastic game under consideration, i.e.,

$$\sup_\pi \phi(x, \pi, g) \leq c, \qquad x \in X, \quad (4.5)$$

and, consequently,

$$\inf_\gamma \sup_\pi \phi(x, \pi, \gamma) \leq c, \qquad x \in X. \quad (4.6)$$

Similarly, fixing the strategy f for the coalition K_1 and considering an appropriate R-model, we obtain

$$\inf_\gamma \phi(x, f, \gamma) \geq c, \qquad x \in X. \quad (4.7)$$

and, consequently,

$$\sup_{\pi} \inf_{\gamma} \phi(x, \pi, \gamma) \geq c, \qquad x \in X. \tag{4.8}$$

From (4.6), (4.8), and inequality

$$\inf_{\gamma} \sup_{\pi} \phi(x, \pi, \gamma) \geq \sup_{\pi} \inf_{\gamma} \phi(x, \pi, \gamma), \qquad x \in X,$$

we have

$$\inf_{\gamma} \sup_{\pi} \phi(x, \pi, \gamma) \equiv \sup_{\pi} \inf_{\gamma} \phi(x, \pi, \gamma) \equiv c.$$

We will prove optimality of strategy f for coalition K_1. (4.3) and (4.4) imply that

$$c + v(x) = r\Big(x, f(\tilde{x}_{V_1}), g(\tilde{x}_{V_2})\Big) + \int v(y) Q\Big(dy \mid x, f(\tilde{x}_{V_1}), g(\tilde{x}_{V_2})\Big),$$
$$x \in X. \tag{4.9}$$

We argue now as in the proof of Theorem 3.51 and obtain from (4.9) $\phi(x, f, g) \equiv c$. From this we see that in (4.5) and (4.7), equality holds. Therefore, for any strategy $\gamma' \in LS$ of coalition K_2

$$\phi(x, f, \gamma') \geq \inf_{\gamma} \phi(x, f, \gamma)$$

$$= c$$

$$= \inf_{\gamma} \sup_{\pi} \phi(x, \pi, \gamma), \qquad x \in X.$$

Optimality of strategy g for coalition K_2 is proved similarly.   □

For every function $u(x)$ in Banach space $M(X)$ of bounded measurable functions on X we denote for all $x \in X$, $\zeta \in \mathsf{P}_{A_{V_1}}$, $\lambda \in \mathsf{P}_{A_{V_2}}$

$$K_u(x, \zeta, \lambda) = r(x, \zeta, \lambda) + \int u(y) Q(dy \mid x, \zeta, \lambda).$$

Theorem 4.10. Let there exist a constant c and a function $v \in M(X)$ such that

$$c + v(x) = \sup_{\zeta \in P_{A_{V_1}}} \inf_{\lambda \in P_{A_{V_2}}} K_v(x, \zeta, \lambda), \qquad x \in X, \qquad (4.10)$$

and assume further that the map F_v, defined by

$$F_v(x) = \left\{ (\zeta', \lambda') : \inf_{\lambda \in P_{A_{V_2}}} K_v(x, \zeta', \lambda) = \sup_{\zeta \in P_{A_{V_1}}} K_v(x, \zeta, \lambda') \right\},$$

is well defined, i.e., $F_v(x) \neq \emptyset$ and closed in $P_{A_1} \times P_{A_2}$, for all $x \in X$, and Borel measurable in the sense of Definition 2.34.

Then the stochastic game has a value that is constant c and both coalitions K_1 and K_2 have optimal stationary local strategies. $\odot$

Proof: Borel measurability of F_v by the Theorem of Choice (Selection Theorem 2.35) guarantees the existence of Borel functions $f \colon \widetilde{X}_{V_1} \to P_{A_{V_1}}$ and $g \colon \widetilde{X}_{V_2} \to P_{A_{V_2}}$ for which

$$\inf_{\lambda \in P_{A_{V_2}}} K_v\big(x, f(\tilde{x}_{V_1}), \lambda\big) = \sup_{\zeta \in P_{A_{V_1}}} K_v\big(x, \zeta, g(\tilde{x}_{V_2})\big), \qquad x \in X.$$

From this we obtain

$$\sup_{\zeta \in P_{A_{V_1}}} \inf_{\lambda \in P_{A_{V_2}}} K_v(x, \zeta, \lambda) \geq \inf_{\lambda \in P_{A_{V_2}}} K_v\big(x, f(\tilde{x}_{V_1}), \lambda\big)$$

$$= \sup_{\zeta \in P_{A_{V_1}}} K_v\big(x, \zeta, g(\tilde{x}_{V_2})\big)$$

$$\geq \inf_{\lambda \in P_{A_{V_2}}} \sup_{\zeta \in P_{A_{V_1}}} K_v(x, \zeta, \lambda), \qquad x \in X.$$

$$(4.11)$$

Because

$$\inf_{\lambda \in P_{A_{V_2}}} \sup_{\zeta \in P_{A_{V_1}}} K_v(x, \zeta, \lambda) \geq \sup_{\zeta \in P_{A_{V_1}}} \inf_{\lambda \in P_{A_{V_2}}} K_v(x, \zeta, \lambda), \qquad x \in X,$$

all inequalities in (4.11) are replaced by equalities. Taking into account (4.10), we obtain (4.3) and (4.4), and by Theorem 4.9, the proof is completed. □

Remark 4.11. It is easy to see that for countable phase space X, the Borel measurability of the map F_v can be replaced by the condition that for all $x \in X$ the sets $F_v(x)$ are nonempty. ⊙

The proofs of the following lemmas are borrowed from the proofs of Lemmas 2.1 and 3.1 in [MP70] for the case of two-person games. They will enable us to prove and apply smoothness properties in our general setting.

Lemma 4.12. Let $p(x, a)$ be a continuous real-valued function on $X \times A$, where X and A are compact spaces. Then for all $x \in X$, $\zeta \in \mathsf{P}_{A_{V_1}}$, $\lambda \in \mathsf{P}_{A_{V_2}}$ the function $p(x, \zeta, \lambda) := \iint p(x, a) \, d\zeta(a_{V_1}) \, d\lambda(a_{V_2})$ is continuous on $X \times \mathsf{P}_{A_{V_1}} \times \mathsf{P}_{A_{V_2}}$. ⊙

Proof: Let $x^n \to x^0$ in X, $\zeta^n \to \zeta^0$ in $\mathsf{P}_{A_{V_1}}$ and $\lambda^n \to \lambda^0$ in $\mathsf{P}_{A_{V_2}}$; and let $\epsilon > 0$. We have

$$\left| p(x^n, \zeta^n, \lambda^n) - p(x^0, \zeta^0, \lambda^0) \right|$$

$$\leq \left| \iint p(x^n, a) \, d\zeta^n(a_{V_1}) \, d\lambda^n(a_{V_2}) - \right.$$

$$\left. - \iint p(x^0, a) \, d\zeta^n(a_{V_1}) \, d\lambda^n(a_{V_2}) \right| +$$

$$+ \left| \iint p(x^0, a) \, d\zeta^n(a_{V_1}) \, d\lambda^n(a_{V_2}) - \right.$$

$$\left. - \iint p(x^0, a) \, d\zeta^0(a_{V_1}) \, d\lambda^0(a_{V_2}) \right|. \quad (4.12)$$

As $X \times A$ is a compact space $p(x, a)$ is uniformly continuous. Hence, there exists N_1 such that whenever $n \geq N_1$ we have $\left| p(x^n, a) - p(x^0, a) \right| < \epsilon/2$ for all $a \in A$. Consequently, the first term on the right-hand side of (4.12) is at most $\epsilon/2$ for all $n \geq N_1$. Next, the product measures $\zeta^n \times \lambda^n$ on A converge weakly to $\zeta^0 \times \lambda^0$. To see this, use the following facts: (a) every real-valued continuous function f on A can be approximated uniformly by a sequence of function of the form $\sum_{i=1}^{k} f_i g_i$, where f_i are continuous, real-valued functions on A_{V_1} and g_i are continuous, real-valued functions on A_{V_2} (Stone–Weierstrass theorem); (b) if f, g are continuous real-valued functions on A_{V_1}, A_{V_2}, respectively, then

$$\iint f\left(a_{V_1}\right) g\left(a_{V_2}\right) d\zeta^n\left(a_{V_1}\right) d\lambda^n\left(a_{V_2}\right)$$

$$\longrightarrow \iint f\left(a_{V_1}\right) g\left(a_{V_2}\right) d\zeta^0\left(a_{V_1}\right) d\lambda^0\left(a_{V_2}\right).$$

Since $p(x^0, \cdot)$ is a continuous function on A henceforth there exists N_2, such that wherever $n \geq N_2$ the second term on the right-hand side of (4.12) is at most $\epsilon/2$. Since ϵ is arbitrary this proves that $p(x^n, \zeta^n, \lambda^n) \to p(x^0, \zeta^0, \lambda^0)$. □

Lemma 4.13. Assume X, $P_{A_{V_1}}$, $P_{A_{V_2}}$ are compact spaces, and let u be a bounded, continuous function on $X \times P_{A_{V_1}} \times P_{A_{V_2}}$. Then the functions $u_*(x) = \max_{\zeta \in P_{A_{V_1}}} \min_{\lambda \in P_{A_{V_2}}} u(x, \zeta, \lambda)$ and $u^*(x) = \min_{\lambda \in P_{A_{V_2}}} \max_{\zeta \in P_{A_{V_1}}} u(x, \zeta, \lambda)$ are continuous. ⊙

Proof: At first, we shall prove that function $u'(x, \lambda) = \max_{\zeta \in P_{A_{V_1}}} u(x, \zeta, \lambda)$ is continuous. Let $(x^n, \lambda^n) \to (x^0, \lambda^0)$ in $X \times P_{A_{V_2}}$. Since $\{u'(x^n, \lambda^n), n \geq 1\}$ is a bounded sequence of real numbers, it has a convergent subsequence. We will prove that every convergent subsequence of $\{u'(x^n, \lambda^n), n \geq 1\}$ goes to $u'(x^0, \lambda^0)$. Assume then that $\{u'(x^{n'}, \lambda^{n'})\}$

is a convergent subsequence. Then there exists $\zeta^{n'} \in \mathsf{P}_{A_{V_1}}$ such that $u'(x^{n'}, \lambda^{n'}) = u(x^{n'}, \zeta^{n'}, \lambda^{n'})$. As $\mathsf{P}_{A_{V_1}}$ is compact, there exists subsequence $\{\zeta^{n''}\}$ of $\{\zeta^{n'}\}$ such that $\zeta^{n''} \to \zeta^0$ (say). It follows from the continuity of u that $u(x^{n''}, \zeta^{n''}, \lambda^{n''}) \to u(x^0, \zeta^0, \lambda^0)$. Moreover, if $\zeta \in \mathsf{P}_{A_{V_1}}$, $u(x^{n''}, \zeta, \lambda^{n''}) \leq u(x^{n''}, \zeta^{n''}, \lambda^{n''})$ and so $u(x^0, \zeta, \lambda^0) \leq u(x^0, \zeta^0, \lambda^0)$, i.e., $u'(x^0, \lambda^0) = u(x^0, \zeta^0, \lambda^0)$. Hence, $u'(x^{n''}, \lambda^{n''}) \to u'(x^0, \lambda^0)$ and so $u'(x^{n'}, \lambda^{n'}) \to u'(x^0, \lambda^0)$. This proves that $u'(x^n, \lambda^n) \to u'(x^0, \lambda^0)$.

Similarly, the function $u''(x, \zeta) = \min_{\lambda \in \mathsf{P}_{A_{V_2}}} u(x, \zeta, \lambda)$ is continuous and, therefore, the functions $u_\star(x) = \max_{\zeta \in \mathsf{P}_{A_{V_1}}} \min_{\lambda \in \mathsf{P}_{A_{V_2}}} u(x, \zeta, \lambda)$ and $u^\star(x) = \min_{\lambda \in \mathsf{P}_{A_{V_2}}} \max_{\zeta \in \mathsf{P}_{A_{V_1}}} u(x, \zeta, \lambda)$ are continuous. $\square$

4.2.2 Existence of optimal Markovian policies and values of the game

We have now prepared the ground to state and prove our main theorems. We need the following standard regularity assumptions for the transition kernels, which will provide conditions to prove optimality properties of local strategies. The proofs of the theorems utilize ideas from the proofs of Theorems 3 and 4 in [Gub72]. Foundations for the development in that paper and technical details can be found in [GS72b].

Assumption 4.14. $Q = \prod_{j \in V} Q_j$ is a local and synchronous Markov kernel according to Definition 4.4 for (ξ, π, γ) and the admissible control sets are independent of time:

$$A(x) := A^t(x) = \underset{i \in V}{\times} A_i^t\left(x_{\widetilde{N}(i)}\right) =: \underset{i \in V}{\times} A_i\left(x_{\widetilde{N}(i)}\right).$$

There exists a non-negative measure μ on $(X, \mathfrak{X})$ with $\mu(X) > 0$ such that:

$$\mu(C) \leq Q(C \mid x, a), \qquad C \in \mathfrak{X}, \quad (x, a) \in \Delta. \qquad \odot$$

Assumption 4.15. $Q = \prod_{j \in V} Q_j$ is a local and synchronous Markov kernel according to Definition 4.4 for (ξ, π, γ) and the admissible control sets are independent of time:

$$A(x) := A^t(x) = \underset{i \in V}{\times} A_i^t \left(x_{\widetilde{N}(i)} \right) =: \underset{i \in V}{\times} A_i \left(x_{\widetilde{N}(i)} \right).$$

There exists a non-negative measure η on $(X, \mathfrak{X})$ with $\eta(X) < 2$ such that:

$$\eta(C) \geq Q(C \mid x, a), \qquad C \in \mathfrak{X}, \quad (x, a) \in \Delta. \qquad \odot$$

Theorem 4.16. Assume that X, A are compact spaces, $r(x, a)$ is continuous on $X \times A$, and the transition probability $Q(\cdot \mid x, a)$ is weakly continuous as a function on $X \times A$. If Assumption 4.14 or 4.15 holds, then the stochastic game has a value which is constant and coalitions K_1 and K_2 have optimal stationary local (Markovian) strategies. $\qquad \odot$

Proof: Let Assumption 4.14 be fulfilled. Define an operator $\mathfrak{U}$ on the Banach space $C(X)$ of continuous bounded functions on X by

$$\mathfrak{U}u(x) = \sup_{\zeta \in \mathsf{P}_{A_{V_1}}} \inf_{\lambda \in \mathsf{P}_{A_{V_2}}} K_u(x, \zeta, \lambda) - \int u(y) \, d\mu(y). \qquad (4.13)$$

Lemmas 4.12 and 4.13 imply that the functions $K_u(x, \zeta, \lambda)$ on $X \times \mathsf{P}_{A_{V_1}} \times \mathsf{P}_{A_{V_2}}$,

and $\sup_{\zeta \in \mathsf{P}_{A_{V_1}}} \inf_{\lambda \in \mathsf{P}_{A_{V_2}}} K_u(x, \zeta, \lambda)$ on X are continuous for every function $u \in C(X)$ and, consequently, sup and inf can be replaced by max and min respectively, ($\mathsf{P}_{A_{V_1}}$ and $\mathsf{P}_{A_{V_2}}$ are compact). Using Assumption 4.14 as in Theorem 3.57, it is easy to check that the operator $\mathfrak{U}$ is a contraction with contraction factor $1 - \mu(X)$ ($0 \leq 1 - \mu(X) < 1$). From Banach's Fixed Point Theorem, there exists in $C(X)$ a function $u^\star(x)$ such that $\mathfrak{U}u^\star = u^\star$. From (4.13), we obtain relation (4.10) with $v(x) = u^\star(x)$ and $c = \int u^\star(y) \, d\mu(y)$.

We shall show that the map $F_{u^\star} : X \to (2)_{set}^{\mathsf{P}_{A_{V_1}} \times \mathsf{P}_{A_{V_2}}}$ defined by (4.10) is upper semicontinuous and, therefore, measurable. For all $x \in X$ the function $K_{u^\star}(x, \zeta, \lambda)$ on $\mathsf{P}_{A_{V_1}} \times \mathsf{P}_{A_{V_2}}$ satisfies all conditions of Minimax Theorem (see Theorem 1 of [Ky53]) and, consequently,

$$\max_{\zeta \in \mathsf{P}_{A_{V_1}}} \min_{\lambda \in \mathsf{P}_{A_{V_2}}} K_{u^\star}(x, \zeta, \lambda) = \min_{\lambda \in \mathsf{P}_{A_{V_2}}} \max_{\zeta \in \mathsf{P}_{A_{V_1}}} K_{u^\star}(x, \zeta, \lambda). \qquad (4.14)$$

Equation (4.14), together with the continuity of the functions $\max_{\zeta \in \mathsf{P}_{A_{V_1}}} K_{u^\star}(x, \zeta, \lambda)$ and $\min_{\lambda \in \mathsf{P}_{A_{V_2}}} K_{u^\star}(x, \zeta, \lambda)$ (see the proof of Lemma 4.13), imply that the sets $F_{u^\star}(x) \subseteq \mathsf{P}_{A_{V_1}} \times \mathsf{P}_{A_{V_2}}$ are nonempty and closed in the product topology of $\mathsf{P}_{A_{V_1}} \times \mathsf{P}_{A_{V_2}}$. To prove semicontinuity of the map $F_{u^\star}$ it is enough to show that

$$\lim_{n \to \infty} (\zeta^n, \lambda^n) = (\zeta', \lambda'), \quad \lim_{n \to \infty} x^n = x, \quad (\zeta^n, \lambda^n) \in F_{u^\star}(x^n) \qquad (4.15)$$

implies

$$(\zeta', \lambda') \in F_{u^\star}(x). \qquad (4.16)$$

From (4.15) we have

$$\max_{\zeta \in \mathsf{P}_{A_{V_1}}} K_{u^\star}(x^n, \zeta, \lambda^n) = \min_{\lambda \in \mathsf{P}_{A_{V_2}}} K_{u^\star}(x^n, \zeta^n, \lambda). \qquad (4.17)$$

Passing to the limit as $n \to \infty$ from continuity of both sides of (4.17), we obtain

$$\max_{\zeta \in \mathsf{P}_{A_{V_1}}} K_{u^\star}(x, \zeta, \lambda') = \min_{\lambda \in \mathsf{P}_{A_{V_2}}} K_{u^\star}(x, \zeta', \lambda),$$

which is equivalent to (4.16).

Theorem 4.10 implies that the stochastic game has a constant value and coalitions K_1 and K_2 have optimal stationary local strategies.

When Assumption 4.15 is fulfilled instead of Assumption 4.14, the proof will be carried out similarly using an operator U' operating in $C(X)$ by

$$\mathfrak{U}'u(x) = -\max_{\zeta \in \mathsf{P}_{A_{V_1}}} \min_{\lambda \in \mathsf{P}_{A_{V_2}}} K_u(x, \zeta, \lambda) - \int u(y) \, d\eta(y).$$

The proof is completed. □

Note that the resulting (in general random) strategies of the Theorem 4.16 can be chosen as smooth versions:

By the Theorem of Choice for semicontinuous maps the functions f and g in Theorem 4.10 here can be chosen from Baire class 1 of $\mathsf{P}_{A_{V_i}}$-valued functions on $\widetilde{X}_{V_i}$.

If the state space X is countable, we write $\mathsf{q}(y \mid x, a) = \mathsf{Q}(\{y\} \mid x, a)$, $x, y \in X$, $a \in A$. We have the following theorem.

Theorem 4.17. Assume X is a countable space and A is compact. Assume further that the functions $r(x, a)$ and $\mathsf{q}(y \mid x, a)$ are continuous on A for every fixed $x, y \in X$, and that Assumption 4.15 holds.

Then the stochastic game has a constant value and both coalitions K_1 and K_2 have optimal stationary local strategies. ⊙

Proof: The existence of the constant c and the bounded function $v(x)$ satisfying (4.10) is proved by applying a fixed-point theorem as in Theorem 4.16, replacing space $C(X)$ by the space of bounded sequences.

Assumption 4.15 implies that series $\sum\limits_{y \in X} v(y)\mathsf{q}(y \mid x, a)$ converges uniformly in x, a (Weierstrass criterion). Therefore, in every fixed $x \in X$ function

$$K_v(x, a) = r(x, a) + \sum_{y \in X} \mathsf{q}(y \mid x, a)$$

is continuous on A and by Lemma 4.12, the function

$$K_v(x, \zeta, \lambda) = \iint K_v(x, a) \, d\zeta\left(a_{V_1}\right) d\lambda\left(a_{V_2}\right)$$

is continuous on $\mathsf{P}_{A_{V_1}} \times \mathsf{P}_{A_{V_2}}$. As in Theorem 4.16, this implies nonemptiness (and closeness) of the sets $F_v(x)$ for every $x \in X$. By Theorem 4.10 and Remark 4.11, the proof is completed. □

Remark 4.18. The results formulated in this section remain valid for stochastic games on graphs where the reward function r for each coalition may depend on the local states of all vertices.　　　$\odot$

Chapter 5

LOCAL CONTROL OF CONTINUOUS TIME INTERACTING MARKOV AND SEMI-MARKOV PROCESSES WITH GRAPH STRUCTURED STATE SPACE

In this chapter we develop models for spatially distributed systems that evolve in continuous time and that are controlled by decision makers who act locally and only have information at hand about the system's state in the neighborhood around their position. As in Chapter 3 for the discrete time networks, we are especially interested in Markovian systems.

This is in line with the main directions of research in the literature: Optimal (sequential) control of stochastic systems is in most cases described by Markov process models. But there is a rich literature on the control of processes that do not show the memoryless property of the holding times in Markovian systems. Introducing nonexponential or nongeometrical sojourn times of the processes leads to considering semi-Markov processes and Markov renewal processes. The first section of this chapter is dedicated to optimization problems in this framework.

5.1 LOCAL CONTROL OF INTERACTING SEMI-MARKOV PROCESSES WITH COMPACT STATE AND ACTION SPACE

Early papers on stochastic dynamic optimization for semi-Markov processes (sometimes called renewal reward processes) are [Jew63a, Jew63b, How64]. For further references to early works, see [Hin70]. More recent work on semi-Markov decision processes can be found, e.g., in [Kit87, VA93, Wak87, LVRA94, Can84], and the references given therein. The general construction of semi-regenerative control processes and semi-Markov control processes with application to queueing systems can be found in the recent book of Kitaev and Rykov [KR95].

There has been a revival of semi-Markov decision processes and renewal reward processes in the area of *performability*, where *performance analysis* and *reliability* of complex systems are investigated in a unified setting over the last twenty years. One of the often applied techniques in the field is to formulate the systems as a Markov or semi-Markov model with a reward structure and then to numerically evaluate and optimize the parameters that are subject to possible control. For an overview and a review of recent results, see [HMRT01].

Introductory definitions of semi-Markov processes and Markov renewal processes as we shall consider in this paper and fundamental investigations on the subject may be found in [Čin75, DK87], and with a special emphasis on optimization in [KR95].

In this section we consider stochastic processes in continuous time with compact state spaces that are structured by an underlying graph as defined in Section 3.1. At any node of the graph there is a local state space, such that the global state space of the process is the product (indexed by the graph) of the local state spaces. The graph then defines a neighborhood structure for the states of the systems. These neighborhoods determine the local interactions of the spatial process coordinates. Then for a fixed time instant the random state of the system, respectively of the describing stochastic process, is a random field with respect to the neighborhood graph.

We assume that the process has the semi-Markov property in time and that the transition kernels for the jumps of the process have a spatial Markov property with respect to the underlying graph.

We consider only step controls as defined in Subsection 2.3.1, page 29. Our aim is to find optimal strategies to control the system when the decision makers located at the nodes of the network can use in decision making only information gathered in their respective neighborhoods. We find conditions under which in the class of local strategies there exist optimal stationary deterministic policies. Our results extend those of [GS72c] and [GS72a] on controlled semi-Markov processes in the direction of interacting semi-Markov processes.

The necessary definitions on semi-Markov processes that evolve in space and time are recalled in Section 5.1.1. We assume time to be continuous but allow decisions only at jump instants of the process (for more details, see [YF79, Yus80, Yus83, Can84]). The random policies show as usual a conditional independence structure. We assume that a similar structure can be found in the transition kernels of the jump chains (synchronized kernels). In Subsection 5.1.3 we first provide abstract conditions for the existence of stationary deterministic strategies in the class of local structures and then show that under some weak smoothness assumptions, these conditions are fulfilled.

5.1.1 Semi-Markov processes with locally interacting components

For systems with locally interacting coordinates we introduced in Section 3.1 a formal definition of an the interaction structure and a state space that is spatially structured by the so defined interaction graph. We assume here that the local state spaces X_i are compact metric spaces with countable basis, endowed with Borel-σ-algebra $\mathfrak{X}_i$. Then $X := \underset{i \in V}{\times} X_i$ is

the global state space of the system, endowed with the product-σ-algebra
$$\mathfrak{X}_V = \sigma\left\{ \underset{i \in V}{\times}\, \mathfrak{X}_i \right\} =: \mathfrak{X}.$$
The evolution over time of our system that has state space $(X, \mathfrak{X}) = \left(\underset{i \in V}{\times}\, X_i, \sigma\left\{ \underset{i \in V}{\times}\, \mathfrak{X}_i \right\} \right)$ is described by a stochastic process $\eta = (\eta^t)$ with random fields as one-dimensional marginals in time.

We assume that time is continuous, $t \in \mathbb{R}_+ = [0, +\infty)$. The subscript k in η_k^t refers to the vertex k, η_k^t therefore denotes the marginal distribution for time t and node k of some vector valued process $\eta = (\eta^t : t \geq 0)$. Such processes that vary in space and time, with the space variable being structured by some graph, were investigated in Chapter 3, using Markov chain techniques. It is our aim here to remove the assumption that in time, the holding times of the processes are memoryless (exponentially or geometrically distributed); i.e., we investigate semi-Markov processes and Markov renewal processes.

Definition 5.1 (Markov renewal process, semi-Markov process).
Consider a stochastic jump process $\eta = (\eta^t : t \geq 0)$ with state space X and paths that are right continuous and have left-hand limits (cadlag paths) defined by a sequence $(\xi, \tau) = \{(\xi^n, \tau^n), n = 0, 1, \dots \}$. The $\mathbb{R}_+$-valued random variables τ^n are the interjump times and the sequence $\{\xi^n, n = 0, 1, \dots\}$ is the sequence of states of the process entered just after the jump instants. To be more precise:

Let $\sigma = \{\sigma^n : n = 0, 1, \dots\}$ be given as $\sigma^0 = 0$, and $\sigma^n = \sum_{i=0}^{n-1} \tau^i$, $n > 0$, the increasing sequence of jump times. Then if $t \in [\sigma^n, \sigma^{n+1})$, we have $\eta^t = \xi^n$, $n \in \mathbb{N}$.

If the (for any n) $\Pr^{(\xi^k, k=0,1,\dots,n-1, \sigma^0, \tau^k, k=0,1,\dots,n-2)}$-almost surely defined conditional distributions of $(\xi, \tau) = \{(\xi^n, \tau^n), n = 0, 1, \dots \}$ fulfill

$$\Pr\{\xi^{n+1} \in C, \tau^n \leq s \mid \xi^k = x^k, k = 0, 1, \dots, n, \dots$$
$$\dots, \sigma^0 = 0, \tau^k = s^k, k = 0, 1, \dots, n-1\}$$
$$= \Pr\{\xi^{n+1} \in C, \tau^n \leq s \mid \xi^n = x^n\}, \qquad n > 0, \quad (5.1)$$

$\Pr^{(\xi^k, k=0,1,\ldots,n-1, \sigma^0, \tau^k, k=0,1,\ldots,n-2)}$-almost surely, and are independent of n, then (ξ, τ) is called a homogeneous Markov renewal process.

We assume that the right side of (5.1) is determined by a transition kernel

$$\widetilde{Q}(C, s \mid x): X \times \mathfrak{X} \times \mathbb{B}_+ \rightarrow [0; 1],$$

where we write

$$\widetilde{Q}(C, s \mid x) := \widetilde{Q}(C \times [0, s] \mid x).$$

The process $\eta = \left(\eta^t \colon 0 \leq t < \sum_{i=0}^{\infty} \tau^i\right)$ is called homogeneous semi-Markov process. The transition function

$$\widetilde{Q}(C, s \mid x) = \Pr\{\xi^{n+1} \in C, \tau^n \leq s \mid \xi^n = x\}$$

is called the semi-Markov kernel of both, the Markov renewal process and the semi-Markov process. We shall henceforth assume that

$$\widetilde{Q}(C, 0 \mid x) = \mathbb{1}_C(x) \quad \forall \, x, C, \quad \text{and} \quad \sum_{i=0}^{\infty} \tau^i = \infty$$

holds and that the sequence $\sigma = \{\sigma^n \colon n = 0, 1, \ldots\}$ has no finite accumulation points. From (5.1) it follows that $\xi = (\xi^n \colon n \in \mathbb{N})$ is a homogeneous Markov chain. The transition probability of this embedded Markov chain is

$$Q(C \mid x) = \widetilde{Q}(C, \infty \mid x) = \Pr\{\xi^{n+1} \in C, \tau^n \leq \infty \mid \xi^n = x\}$$
$$\forall \, x \in X, \quad C \in \mathfrak{X}.$$

We assume a further transition kernel being given that determines the conditional sojourn times of the process given the present and the next state of the system. This we denote by

$$\mathrm{T}(\cdot \mid x, y) = \Pr\{\tau^n \leq \cdot \mid \xi^n = x, \xi^{n+1} = y\}, \quad x, y \in X \qquad (5.2)$$

where on the right side the conditional sojourn time distribution of the sojourn time τ^n given the process is in state x and will be next in y when τ^n has expired is only $\Pr^{\xi^n, \xi^{n+1}}$-almost surely defined. $\odot$

5.1.2 Controlled semi-Markov processes with local and synchronous transition mechanisms

The aim of this section is to introduce a controlled semi-Markov process with locally interacting synchronous components similar to that constructed in the pure Markovian framework in Chapter 3 as controlled Markov random field.

   If a semi-Markov process η and its associated Markov renewal process (ξ, τ) from their very construction are related to the neighborhood system $\{N(k) \colon k \in V\}$ of (V, B), it is natural to assume that with respect to the evolution over time, the value $\xi_k^n = \eta_k(\sigma^n) = x_k$ of the k-th vertex depends on the previous states of the whole system only through the values of the vertices in $\widetilde{N}(k)$ (including k) after transition moment σ^{n-1}. To describe this, we introduce the Markov property in space for the kernel of the embedded jump chain for a semi-Markov process along the lines of Definition 3.10. Because we concentrate on the local behavior of the embedded jump chain, we arrive at just the definition of synchronous and local kernels as given there. It follows that we can express the required properties directly in terms of the transition kernel of the embedded Markov chain $\mathbf{Q}$. For the reader's convenience, we recall the relevant definitions.

Definition 5.2 (Synchronized and local transition kernels). Let $\eta = (\eta^t \colon t \geq 0)$ with state space X be a semi-Markov process with associated Markov renewal process $(\xi, \tau) = \{(\xi^n, \tau^n), n = 0, 1, \dots\}$. The jump transition probabilities of η and ξ are said to be *local* [Vas78, p. 100] if for all $k \in V$, $x^0, \dots, x^{n+1} \in X$, $C_k \in \mathfrak{X}_k$

$$\Pr\{\xi_k^{n+1} \in C_k \mid \xi^n = x^n, \dots, \xi^0 = x^0\}$$
$$= \Pr\left\{\xi_k^{n+1} \in C_k \mid \xi_{\widetilde{N}(k)}^n = x_{\widetilde{N}(k)}^n\right\} \qquad (5.3)$$

holds, i.e., the state of vertex k just after the jump depends on the state of its complete neighborhood at the previous transition instant only. The transition probabilities of X are said to be *synchronous* [Vas78, p. 100] if

$$\Pr\{\xi_K^{n+1} \in C_K \mid \xi^n = x^n\} \tag{5.4}$$
$$= \prod_{k\in K} \Pr\{\xi_k^{n+1} \in C_k \mid \xi^n = x^n\}, \qquad \forall\, K \subset V, \quad x^n \in X,$$

where $C_K = \underset{k\in K}{\times}\, C_k \in \mathfrak{X}_K$.

If η, ξ, or (ξ, τ), fulfill (5.3) and (5.4), then η is called a *semi-Markov process with locally interacting synchronous components over* (Γ, X), shortly, a *semi-Markov random field*. Similarly we call (ξ, τ) a *Markov renewal process with locally interacting synchronous components over* (Γ, X), shortly, a *Markov renewal field*. $\qquad \odot$

We introduce classes of admissible policies to control the interacting coordinates of the stochastic processes under consideration. For simplicity of the presentation we assume that some (uncontrolled) semi-Markov random field $\eta = (\eta^t \colon t \geq 0)$ or $(\xi, \tau) = \{(\xi^n, \tau^n) \colon n \in \mathbb{N}\}$ according to Definition 5.1 is given as our starting point. η will be equipped in the sequel with a control structure that governs the jump transition kernel and the sojourn time distributions of (ξ, τ). We mainly consider controls, which are local in the sense to be specified now. The definition is in parallel to Definition 3.16.

Definition 5.3 (Action spaces and local restrictions). The sequence of decision instants (control instants) is $\sigma = \{\sigma^n, n = 0, 1, \ldots\}$.

(1) The set of actions (control values) usable at control instants is $A = \underset{i\in V}{\times}\, A_i$ over Γ, where A_i is a set of possible actions (decisions) for vertex i. We assume that A_i is a compact metric space with countable basis and with Borel-σ-algebra $\mathfrak{A}_i$. $\mathfrak{A}$ is the Borel-σ-algebra over A.

(2) If for the decision maker at node i at time σ^n under state $\xi^n = x$ the set of control actions is restricted to $A_i^n(x) \subset A_i$, we call $A_i^n(x)$ the set of admissible actions (decisions) at time σ^n in state x. We always assume that the restriction sets are time invariant and therefore depend on the actual state of the system only. We denote $A_i(x) := A_i^n(x) \subset A_i$, for all decision instants σ^n.

(3) We assume that the so defined set valued maps

$$A_i \colon X \to 2^{A_i} - \{\emptyset\}, \qquad x \to A_i(x), \qquad i \in V,$$

depend on $x \in X$ only through $x_{\widetilde{N}(i)} \in X_{\widetilde{N}(i)}$ and are Borel measurable in the sense of Definition 2.34. Thus the A_i determine admissibility as a local property. We therefore write $A_i(x) =: A_i\left(x_{\widetilde{N}(i)}\right)$. We also assume the sets

$$\kappa_i = \left\{ \left(x_{\widetilde{N}(i)}, a_i\right) : x_{\widetilde{N}(i)} \in X_{\widetilde{N}(i)}, a_i \in A_i\left(x_{\widetilde{N}(i)}\right) \right\}$$

to be Borel measurable sets of the product space $X_{\widetilde{N}(i)} \times A_i$ and $\kappa = \underset{i \in V}{\times} \kappa_i$ to be Borel measurable in the product space $\widetilde{X} \times A$, $\widetilde{X} = \underset{i \in V}{\times} X_{\widetilde{N}(i)}$ and

$$\widetilde{\mathfrak{X}} = \sigma\left\{ \underset{i \in V}{\times} \mathfrak{X}_{\widetilde{N}(i)} \right\}. \hspace{4cm} \odot$$

Remark 5.4. We sometimes refer to decisions without specifying a local structure. Then we have a global mapping

$$A \colon X \to 2^A, \qquad x \to A(x) - \{\emptyset\},$$

and we define

$$\bar{\kappa} := \left\{ (x, a) \in X \times A : a \in A(x) \right\}. \hspace{2.5cm} \odot$$

Remark 5.5. Cantaluppi [Can84, Theorem 15] proved that in case of infinite time horizon, there always exist optimal policies that use as feasible decision instants the sequence $\sigma = \{\sigma^n, n = 0, 1, \ldots\}$. $\hspace{1cm} \odot$

Due to the random holding times in semi-Markovian decision processes, we have to include the past sojourn time durations into the system's
history. We allow actions of the decision makers only after the transition
instants of the process when the state has just changed. Then the decisions will in general open the possibility to control the jump probabilities
into the next state as well as the duration of the next sojourn time. This
implies that we have to adapt the notion of strategies to the setting of
semi-Markov processes as well.

Definition 5.6 (Histories for semi-Markov processes). A sequence
$(\bar{H}^n \colon n \in \mathbb{N})$ of histories for a semi-Markov decision model where decisions
are only allowed just after jump instants is given by $\bar{H}^0 = X$, $\bar{H}^{n+1} =$
$\bar{H}^n \times A \times \mathbb{R}_+ \times X$ for $n \geq 0$. (The real valued coordinates of the histories
describe the inter-jump times.) Each $\bar{H}^n$, which contains $3n+1$ factor sets,
is endowed with the respective product σ-algebra $\bar{\mathfrak{H}}^n$.

The sequence $A = (A^n \colon n \in \mathbb{N})$ of set valued functions determines
the admissible actions $A^n \colon H^n \subseteq \bar{H}^n \to 2^A - \{\emptyset\}$, and the domain H^n is
recursively defined as $H^0 := X$, and $H^{n+1} := \{(h, a, t, x) \in \bar{H}^{n+1} \colon h \in$
$H^n, a \in A^n(h), t \in \mathbb{R}_+, x \in X\}$. $A^n(h)$ is the set of admissible actions at
time n under history h.

H^n, is endowed with the trace-σ-algebra $\mathfrak{H}^n := H^n \cap \bar{\mathfrak{H}}^n$.

We denote $K^n := \{(h, a) \colon h \in H^n, a \in A^n(h)\}$, and shall always
assume that these sets contain the graph of a measurable mapping. K^n is
endowed with the trace of the product-σ-algebra $\mathfrak{K}^n := K^n \cap \bar{\mathfrak{H}}^n \times \mathfrak{A}$. ⊙

Definition 5.7 (Strategies). (1) Let α_i^n denote the action chosen by the
decision maker at node i at the n-th transition instant σ^n, $\alpha^n := (\alpha_i^n \colon i \in$
$V)$ the joint decision vector at time $\sigma^n, n \in \mathbb{N}$.

(2) A strategy (policy) π to control the system with interacting components is defined as vector of coordinate policies $\pi = (\pi_i, i \in V)$, where
for node i $\pi_i = \{\pi_i^0, \dots, \pi_i^n, \dots\}$ is a sequence of transition probabilities
$\pi_i^n = \pi_i^n(\cdot \mid x^0, a^0, t^0, \dots, x^{n-1}, a^{n-1}, t^{n-1}, x^n)$. So π_i^n is a probability

measure on $(A_i, \mathfrak{A}_i)$ for any $\left(x^0, a^0, t^0, \ldots, x^{n-1}, a^{n-1}, t^{n-1}, x^n\right)$ and measurably dependent on the history $h^n = \left(x^0, a^0, t^0, \ldots, x^{n-1}, a^{n-1}, t^{n-1}, x^n\right)$ of the system up to n-th transition. We therefore have for all $B_i \in \mathfrak{A}_i$

$$\Pr\{\alpha_i^n \in B_i \mid \xi^0 = x^0, \alpha^0 = a^0, \tau^0 = t^0, \ldots,$$
$$\ldots, \xi^{n-1} = x^{n-1}, \alpha^{n-1} = a^{n-1}, \tau^{n-1} = t^{n-1}, \xi^n = x^n\}$$
$$= \pi_i^n \left(B_i \mid x^0, a^0, t^0, \ldots, x^{n-1}, a^{n-1}, x^n\right).$$

(3) In parallel to the synchronous transitions and the locality of the transition kernels, we always assume that the decision makers located at the nodes act conditionally independent given the history of the system. This leads to control of the process governed by a synchronous control kernel

$$\Pr\Big\{\alpha^n \in \underset{i \in V}{\times} B_i \mid \xi^0 = x^0, \alpha^0 = a^0, \tau^0 = t^0, \ldots$$
$$\ldots, \xi^{n-1} = x^{n-1}, \alpha^{n-1} = a^{n-1}, \tau^{n-1} = t^{n-1}, \xi^n = x^n\Big\}$$
$$= \prod_{i \in V} \Pr\left\{\alpha_i^n \in B_i \mid \xi^0 = x^0, \alpha^0 = a^0, \tau^0 = t^0, \ldots\right.$$
$$\ldots, \xi^{n-1} = x^{n-1}, \alpha^{n-1} = a^{n-1}, \tau^{n-1} = t^{n-1}, \xi^n = x^n\}$$
$$= \prod_{i \in V} \pi_i^n \left(B_i \mid x^0, a^0, t^0, \ldots, x^{n-1}, a^{n-1}, t^{n-1}, x^n\right),$$
$$B_i \in \mathfrak{A}_i, \quad a^s \in A, \quad x^s \in X, \quad t^s \in \mathbb{R}_+. \quad \odot$$

Definition 5.8 (Local strategies). Recall that we always assume that for actions the restriction sets are time invariant and depend on neighborhoods only:

$$A^n(x) = A(x) = \underset{i \in V}{\times} A_i\left(x_{\widetilde{N}(i)}\right) \qquad \text{for all } n.$$

(1) If at transition times $\sigma^n, n = 0, 1, \ldots$, the decision α_i^n at node i is made according to the probability π_i^n on basis of the local history $h_i^n =$

$\left(x^0_{\widetilde{N}(i)}, a^0_i, t^0, \ldots, x^{n-1}_{\widetilde{N}(i)}, a^{n-1}_i, t^{n-1}, x^n_{\widetilde{N}(i)} \right)$ of the neighborhood $\widetilde{N}(i)$ of i

only, and if $\pi^n_i \left(A_i \left(x^n_{\widetilde{N}(i)} \right) \mid h^n_i \right) = 1$, for $x^s_{\widetilde{N}(i)} \in X_{\widetilde{N}(i)}$, $a^s_i \in A^s_i \left(x^s_{\widetilde{N}(i)} \right)$,

$t^s \in \mathbb{R}_+$, then π^n_i is said to be admissible local, and the sequence of transition probabilities (decisions) $\pi_i = \{ \pi^n_i, n \in \mathbb{N} \}$ is called admissible local strategy for vertex i.

(2) An admissible local strategy $\pi = (\pi_i, i \in V)$ is called admissible local Markov strategy if

$$\pi^n_i \left(\cdot \mid x^0_{\widetilde{N}(i)}, a^0_i, t^0, \ldots, x^{n-1}_{\widetilde{N}(i)}, a^{n-1}_i, t^{n-1}, x^n_{\widetilde{N}(i)} \right) = \pi^n_i \left(\cdot \mid x^n_{\widetilde{N}(i)} \right), \quad i \in V.$$

(3) An admissible local Markov strategy $\pi = (\pi_i, i \in V)$ is called admissible local stationary (Markov) strategy if $\pi^{n'}_i \left(\cdot \mid x_{\widetilde{N}(i)} \right) = \pi^{n''}_i \left(\cdot \mid x_{\widetilde{N}(i)} \right)$, $i \in V$, for all n', n'' and all x.

(4) An admissible local stationary strategy $\pi = (\pi_i, i \in V)$ is called admissible local stationary deterministic (nonrandomized) strategy if $\pi_i \left(\cdot \mid x_{\widetilde{N}(i)} \right)$, $i \in V$, are one-point measures on $A_i \left(x_{\widetilde{N}(i)} \right)$, $i \in V$, accordingly for all $x \in X$.

The class of all admissible local strategies (with time invariant restriction sets) is denoted by LS; the subclass of admissible local Markov strategies by LS_M. By $LS_S \subseteq LS_M$, we denote the class of admissible local stationary strategies. By LS_P, we denote the class of all admissible local deterministic (= pure) strategies (with time invariant restriction sets), and by LS_D, the subclass of admissible local stationary deterministic strategies. $\odot$

We now incorporate into our semi-Markov process framework synchronization and locality of the transition kernel (similar to (5.3) and (5.4)) and of the decision rules, and then the decision dependent transition mechanism and the decision dependent sojourn time behavior of the system.

The law of motion of the system is characterized by a set of time invariant transition probabilities. Whenever the system enters state

$\xi^n = x^n$ and decision $\alpha^n = a^n$ is made, the transition probability is $\Pr\left\{\xi^{n+1} \in C \mid \xi^n = x^n, \alpha^n = a^n\right\} =: Q(C \mid x^n, a^n)$, which is assumed to be independent of the past given the present generalized state, which includes the present actions made. Then, given $\left\{\xi^n = x^n, \alpha^n = a^n\right\}$ the next state $\xi^{n+1} = x^{n+1}$ of the system is sampled according to $Q(\cdot \mid x^n, a^n)$, and thereafter the sojourn time in x^n, given x^n, a^n, x^{n+1}, is sampled according to some distribution function $T(\cdot \mid x^n, a^n, x^{n+1})$, which is Borel measurable on $\bar{\kappa} \times X$.

We further assume that this transition probability and the sojourn time distributions are independent of n, and that the motion is homogeneous in time.

Applying a control policy π to a semi-Markov process with interacting components η (semi-Markov random field), respectively to the associated Markov renewal process (ξ, τ), as defined in Definition 5.2, we shall call the pair (η, π), respectively the triple (ξ, τ, π) a controlled version of η, respectively (ξ, τ), using strategy π.

It should be noted that for such a controlled process, in general, even the embedded jump chain is not Markovian because the sequence (α^n) of decisions according to π_i^n depends not only on states $x_{\widetilde{N}(i)}^n$, $i \in V$, but on the previous (local) states $x_{\widetilde{N}(i)}^0, \ldots, x_{\widetilde{N}(i)}^{n-1}$ as well. An immediate consequence of the definition will be that if a Markov strategy π is applied as control strategy for the semi-Markov random field, then we obtain an embedded Markov jump chain ξ from (η, π), respectively (ξ, τ, π).

Definition 5.9 ([DKC03]). A pair (η, π), respectively a triple (ξ, τ, π), is called a controlled stochastic jump process with locally interacting synchronous components with respect to the finite interaction graph $\Gamma = (V, B)$, if the following holds:

- $\xi = (\xi^n : n \in \mathbb{N})$ is a stochastic process with state space $X = \underset{i \in V}{\times} X_i$,

$\tau = (\tau^n : n \in \mathbb{N})$ is stochastic process with state space $\mathbb{R}_+$;

- $\eta = (\eta^t : t \geq 0)$ is a stochastic process with state space $X = \underset{i \in V}{\times} X_i$,

and these processes are connected pathwise as follows:

The $\mathbb{R}_+$-valued random variables τ^n are the interjump times and the sequence $\{\xi^n, n = 0, 1, \dots\}$, is the sequence of states of the process entered just after the jump instants.

- $\sigma = \{\sigma^n \colon n = 0, 1, \dots\}$ with $\sigma^0 = 0$, and $\sigma^n = \sum_{i=0}^{n-1} \tau^i$, $n > 0$ is the increasing sequence of jump times. Then, if $t \in [\sigma^n, \sigma^{n+1})$, we have $\eta^t = \xi^n$, $n \in \mathbb{N}$.

We assume that the conditional distribution function of τ^n given $\{\xi^{n+1} = x^{n+1}, \xi^n = x^n, \alpha^n = a^n\}$ is $\mathrm{T}(\cdot \mid x^n, a^n, x^{n+1})$ defined as Borel measurable function on $\kappa \times X$. $\pi = (\pi_i \colon i \in V)$ is an admissible local strategy, and the transitions of ξ are determined as follows. The $\mathrm{Pr}^{(\xi^0, \alpha^0, \dots, \xi^{n-1}, \alpha^{n-1}, \xi^n, \alpha^n)}$-almost surely defined conditional probabilities of the semi-Markov process fulfill

$$\mathrm{Pr}\left\{\xi_K^{n+1} \in C_K \mid \xi^0 = x^0, \alpha^0 = a^0, \dots\right.$$
$$\left. \dots, \xi^{n-1} = x^{n-1}, \alpha^{n-1} = a^{n-1}, \xi^n = y, \alpha^n = a\right\}$$
$$= \mathrm{Pr}\left\{\xi_K^{n+1} \in C_K \mid \xi^n = y, \alpha^n = a\right\}$$
$$= \prod_{j \in K} \mathrm{Pr}\left\{\xi_j^{n+1} \in C_j \mid \xi^n = y, \alpha^n = a\right\}$$
$$= \prod_{j \in K} \mathrm{Pr}\left\{\xi_j^{n+1} \in C_j \mid \xi_{\widetilde{N}(j)}^n = y_{\widetilde{N}(j)}, \alpha_j^n = a_j\right\}$$
$$= \prod_{j \in K} \mathrm{Q}_j\left(C_j \mid y_{\widetilde{N}(j)}, a_j\right)$$
$$= \mathrm{Q}_K(C_K \mid y, a), \quad K \subseteq V, \ y \in X, \ a_j \in A_j\left(y_{\widetilde{N}(j)}\right). \quad (5.5)$$

Here

$$\mathrm{Q}_K(\cdot \mid \cdot, \cdot) \colon \bigotimes_{i \in K} \mathfrak{X}_i \times X \times A \to [0, 1], \quad (C_K, y, a) \to \mathrm{Q}_K(C_K \mid y, a)$$

is a transition kernel, and similarly are defined the $\mathrm{Q}_j\left(C_j \mid y_{\widetilde{N}(j)}, a_j\right)$. If $K = V$ we shall write $\mathrm{Q}_V(C_V \mid y, a) = \mathrm{Q}(C \mid y, a)$. In analogy to Definition 5.2 the Markov kernel $\mathrm{Q} = \prod_{j \in V} \mathrm{Q}_j$ is said to be local and synchronous. Combining then the distributions of the interjump times and the jump probabilities with their specific structure, this construction results in a semi-Markov kernel for the controlled process

$$\widetilde{Q}(C, s \mid x, a) = \Pr\left\{\xi^{n+1} \in C, \tau^n \le s \mid \xi^n = x^n, \alpha^n = a^n\right\}.$$

We shall denote

$$\widetilde{Q}(C, \infty \mid x, a) = \Pr\left\{\xi^{n+1} \in C \mid \xi^n = x^n, \alpha^n = a^n\right\} = Q(C \mid x^n, a^n).$$

η, respectively (ξ, τ) will then be called shortly a controlled time dependent semi-Markov random field, or a controlled time dependent Markov renewal field.   ⊙

Comments on (5.5) are mainly similar to those in the case of standard interaction processes and can be found after Definition 3.21. The additional remark to be added here is that the interjump times are determined globally, so the locally determined jumps are strongly coupled in time.

5.1.3 Criteria for optimality

Consider a time dependent semi-Markov random field η. If the k-th state of the system is x^k and the joint decision a^k is made and if the duration of the sojourn time in x^k is t^k, then a random reward $r(t^k, x^k, a^k)$ is earned. The function $r(s, x, a)$ is assumed to be Borel measurable on $[0, +\infty) \times \kappa$.

We shall evaluate the quality, or optimality, of an applied strategy π with respect to the following long time average reward measure:

$$\phi(x^0, \pi) = \liminf_{n \to \infty} \frac{\mathrm{E}_{x^0}^\pi \sum_{k=0}^n r(\tau^k, \xi^k, \alpha^k)}{\mathrm{E}_{x^0}^\pi \sum_{k=0}^n \tau^k}, \qquad (5.6)$$

where $\mathrm{E}_{x^0}^\pi$ is expectation associated with the controlled process (η, π), or (ξ, τ, π), if $\xi^0 = x^0$. The aim of our investigation is to find conditions for the existence of optimal Markovian policies.

Definition 5.10. A strategy $\pi^\star \in LS$ is called optimal with respect to the maximin reward criterion in the class LS of admissible randomized local strategies if

$$\phi(x, \pi^\star) = \sup_{\pi \in LS} \phi(x, \pi), \qquad \forall\, x \in X.$$

Such a $\pi^\star$ will shortly be said to be locally optimal. ⊙

The quality criterion (5.6) was invented by Ross [Ros70] and used in [GS72c] as well. In [Ros70] only stationary policies are allowed and decisions are made on the basis of the present state only. In this case the value of (5.6) depends only on the transition kernels $Q(\cdot \mid x, a)$, $\pi(\cdot \mid x)$ and the conditional expectations

$$\tau(x, a) = \int_X \int_0^\infty t\, d\mathrm{T}(t \mid x, a, y) Q(dy \mid x, a),$$

$$r(x, a) = \int_X \int_0^\infty r(t \mid x, a)\, d\mathrm{T}(t \mid x, a, y) Q(dy \mid x, a).$$

Then one is in a position to restrict proofs to special forms of sojourn time distributions, say one point distributions, given by their distribution functions, or the one-point reward function with respect to the first component of the arguments:

$$\mathrm{T}(t \mid x, a, y) = \begin{cases} 1, & t \geq \tau(x, a); \\ 0, & t < \tau(x, a), \end{cases}$$

$$r(t, x, a) = \begin{cases} 0, & t < \tau(x, a); \\ r(x, a), & t \geq \tau(x, a). \end{cases}$$

Because we allow general policies, such a property does not hold for the asymptotic average reward in general.

5.1.4 Existence of optimal Markov strategies in the class of local strategies

We shall need in the following some technical assumptions, which we collect here. Recall that

$$
\kappa = \underset{i \in V}{\times} \kappa_i = \underset{i \in V}{\times} \left\{ \left(x_{\widetilde{N}(i)}, a_i \right) \colon x_{\widetilde{N}(i)} \in X_{\widetilde{N}(i)}, a_i \in A_i \left(x_{\widetilde{N}(i)} \right) \right\}
$$

and

$$
\bar{\kappa} = \left\{ (x, a) \in X \times A \colon a \in A(x) \right\}.
$$

Assumption 5.11. $\tau(x, a) \geq m > 0, \qquad (x, a) \in \bar{\kappa};$ ⊙

Assumption 5.12. $\tau(x, a) \leq M < \infty, \qquad (x, a) \in \bar{\kappa};$ ⊙

Assumption 5.13. there exists a non-negative measure μ on $(X, \mathfrak{X})$ such that:

a) $\mu(C) \leq Q(C \mid x, a), \quad (x, a) \in \bar{\kappa}, \ C \in \mathfrak{X};$
b) $\mu(X) > 0.$ ⊙

Let us denote by H^n a random history of the system up to time σ^n. So H^n takes values of the form $h^n = \left(x^0, a^0, t^0, \dots, x^{n-1}, a^{n-1}, t^{n-1}, x^n \right)$. Recall that the restriction sets for the local strategies are time independent.

We define the conditional sojourn time expectations

$$
\tau(x^k, a^k) = \mathsf{E}_x^\pi \left\{ \tau^k \mid H^k = h^k, \alpha^k = a^k \right\}
$$
$$
= \int_X \int_0^\infty t \, d\Gamma(t \mid x^k, a^k, y) Q(dy \mid x^k, a^k),
$$

and the conditional reward expectations

$$r\left(x^k, a^k\right) = \mathsf{E}_x^\pi \left\{ r\left(\tau^k, \xi^k, \alpha^k\right) \mid H^k = h^k, \alpha^k = a^k \right\}$$
$$= \int_X \int_0^\infty r\left(t, x^k, a^k\right) d\Gamma\left(t \mid x^k, a^k, y\right) \mathsf{Q}\left(dy \mid x^k, a^k\right).$$

Using this notation, it follows

$$\mathsf{E}_x^\pi \sum_{k=0}^n r\left(\tau^k, \xi^k, \alpha^k\right) = \mathsf{E}_x^\pi \sum_{k=0}^n r\left(\xi^k, \alpha^k\right), \qquad (5.7)$$

and

$$\mathsf{E}_x^\pi \sum_{k=0}^n \tau^k = \mathsf{E}_x^\pi \sum_{k=0}^n \tau^k\left(\xi^k, \alpha^k\right).$$

Theorem 5.14. If Assumption 5.11 holds and if there exist a constant q and a bounded function $v(x)$ on X such that

$$v(x) = \sup_{a \in A(x)} \left\{ r(x, a) + \int v(y) \mathsf{Q}(dy \mid x, a) - q\tau(x, a) \right\}, \quad x \in X. \ (5.8)$$

Then

$$\sup_{\pi \in LS} \phi(x, \pi) \le q, \qquad x \in X. \qquad (5.9)$$

If thus

$$v(x) = \max_{a \in A(x)} \left\{ r(x, a) + \int v(y) \mathsf{Q}(dy \mid x, a) - q\tau(x, a) \right\}, \qquad x \in X,$$

and for some strategy $\pi^\star \in LS_D$

$$v(x) = r\left(x, \pi^\star(x)\right) + \int v(y) \mathsf{Q}\left(dy \mid x, \pi^\star(x)\right) - q\tau\left(x, \pi^\star(x)\right),$$

$$x \in X, \quad (5.10)$$

holds, then $\pi^\star$ is locally optimal and

$$\phi(x, \pi^\star) \equiv q. \qquad \qquad \odot$$

Proof: For any strategy π, we have

$$E_x^\pi \left\{ \sum_{k=1}^n \left[v(\xi^k) - E^\pi \left\{ v(\xi^k) \mid H^{k-1}, \alpha^{k-1} \right\} \right] \right\} = 0. \tag{5.11}$$

From (5.5) and (5.8), we have

$$E^\pi \left\{ v(\xi^k) \mid H^{k-1} = h^{k-1}, \alpha^{k-1} = a^{k-1} \right\}$$

$$= \int v(y) Q(dy \mid x^{k-1}, a^{k-1})$$

$$= r(x^{k-1}, a^{k-1}) + \int v(y) Q(dy \mid x^{k-1}, a^{k-1}) -$$

$$- q\tau(x^{k-1}, a^{k-1}) - r(x^{k-1}, a^{k-1}) + q\tau(x^{k-1}, a^{k-1})$$

$$\leq v(x^{k-1}) - r(x^{k-1}, a^{k-1}) + q\tau(x^{k-1}, a^{k-1}). \tag{5.12}$$

Substituting (5.12) in (5.11), we obtain

$$E_x^\pi \left\{ \sum_{k=1}^n r(\xi^{k-1}, \alpha^{k-1}) + v(\xi^n) - v(x^0) \right\} \leq q E_x^\pi \sum_{k=1}^n \tau(\xi^{k-1}, \alpha^{k-1}). \tag{5.13}$$

From boundedness of $v(x)$, Assumption 5.11, and (5.7), we obtain

$$\phi(x, \pi) \leq q, \qquad x \in X, \tag{5.14}$$

so inequality (5.9) is proved.

To prove the second part of the theorem, we notice that for the strategy π^* in relations (5.12)–(5.14), equality holds. $\qquad\square$

Remark 5.15. The proof of Theorem 5.14 is similar to the proof of Theorem 2 in [Ros70], which deals with the class of stationary strategies only. Further, our Assumption 5.11 is weaker than that used there. $\qquad\odot$

Let $M(X)$ be Banach space of bounded Borel measurable functions on X with sup-norm $\|u\| = \sup_{x \in X} |u(x)|$. Denote by ρ the metric on $M(X)$ induced by that norm. We define an operator $\mathfrak{U}$ on $M(X)$ by

$$\mathfrak{U}u(x) = \sup_{a \in A(x)} \left\{ r(x,a) + \int u(y) \mathsf{Q}'(dy \mid x,a) \right\}, \qquad (5.15)$$

where

$$\mathsf{Q}'(C \mid x,a) = \mathsf{Q}(C \mid x,a) - \frac{1}{M} \mu(C) \tau(x,a). \qquad (5.16)$$

Theorem 5.16. Let Assumptions 5.11–5.13 hold and assume further that

1) the operator $\mathfrak{U}$ maps some metric subspace $S(X) \subseteq M(X)$ (with metric ρ induced by norm of $M(X)$) into itself;

2) for every function $u \in S(X)$ the map $A_u \colon X \to 2^A - \{\emptyset\}$, which determines for every $x \in X$ the set

$$A_u(x) := \left\{ a \colon a \in A(x),\ \mathfrak{U}u(x) = r(x,a) + \int u(y) \mathsf{Q}'(dy \mid x,a) \right\},$$

is Borel measurable.

Then there exists a strategy in LS_D that is (locally) optimal in LS. ⊙

Proof: We will show that the operator $\mathfrak{U}$ has the following properties:

a) $\mathfrak{U}$ is isotone, i.e., if $u_1(x) \geq u_2(x)$, $x \in X$, then $\mathfrak{U}u_1(x) \geq \mathfrak{U}u_2(x)$, $x \in X$;

b) For any non-negative constant c

$$\mathfrak{U}\big(u(x) + c\big) \leq \mathfrak{U}u(x) + \alpha c, \qquad x \in X, \quad \text{with} \quad \alpha = 1 - \frac{m}{M}\mu(X).$$

Indeed, property a) follows directly from Assumption 5.13 a), and property b) is seen as follows:

$$\mathfrak{U}\big(u(x) + c\big)$$

$$= \sup_{a \in A_x} \left\{ r(x, a) + \int u(y) Q'(dy \mid x, a) + c \int Q'(dy \mid x, a) \right\}$$

$$= \sup_{a \in A_x} \left\{ r(x, a) + \int u(y) Q'(dy \mid x, a) + c\left(1 - \frac{1}{M}\tau(x, a)\mu(X)\right) \right\}$$

$$\leq \sup_{a \in A_x} \left\{ r(x, a) + \int u(y) Q'(dy \mid x, a) + c\left(1 - \frac{m}{M}\mu(X)\right) \right\}$$

$$= \mathfrak{U}u(x) + \alpha c.$$

Properties a) and b) guarantee that $\mathfrak{U}$ is a contraction operator with contraction coefficient $\alpha = 1 - \frac{m}{M}\mu(X) \in [0, 1)$, because μ is substochastic and $m \leq M$. Indeed, apply $\mathfrak{U}$ to both sides of the inequality $u_1(x) \leq u_2(x) + \rho(u_1, u_2)$. From a), b) we have

$$\mathfrak{U}u_1(x) \leq \mathfrak{U}\big(u_2(x) + \rho(u_1, u_2)\big) \leq \mathfrak{U}u_2(x) + \alpha\rho(u_1, u_2).$$

Consequently,

$$\mathfrak{U}u_1(x) - \mathfrak{U}u_2(x) \leq \alpha\rho(u_1, u_2).$$

Exchanging u_1 and u_2 we have

$$\big|\mathfrak{U}u_1(x) - \mathfrak{U}u_2(x)\big| \leq \alpha\rho(u_1, u_2), \qquad x \in X,$$

i.e.,

$$\rho(\mathfrak{U}u_1, \mathfrak{U}u_2) \leq \alpha\rho(u_1, u_2).$$

Banach's fixed point theorem then implies that in $S(X)$ there exists a function $u^\star(x)$ such that

$$u^\star(x) = \sup_{a \in A(x)} \left\{ r(x, a) + \int u^\star(y) Q'(dy \mid x, a) \right\}, \qquad x \in X. \quad (5.17)$$

From (5.17) and (5.16) we obtain the Optimality Equation (5.8) with $v(x) = u^{\star}(x)$ and $q = \frac{1}{M} \int u^{\star}(y)\mu(dy)$.

Assumption 2) guarantees that the map $A_{u^{\star}}$ is Borel measurable. This, by Definition 2.34, especially yields $A_{u^{\star}}(x) \neq \emptyset$ for all $x \in X$, and, from the definition, $A_{u^{\star}}(x) \subseteq A(x)$ holds. So, the set $A_{u^{\star}}(x)$ of maximizers of $\mathfrak{U}u^{\star}(x)$ contains only local policies.

Now, by the Theorem of Choice, there exists a Borel measurable $A(x)$-valued selector $\pi^{\star}$; i.e., we have

$$\pi^{\star}(x) \in \left\{ a \in A(x) \colon \mathfrak{U}u^{\star}(x) = r(x,a) + \int u^{\star}(y)Q'(dy \mid x,a) \right\}$$

which means

$$\mathfrak{U}u^{\star}(x) = r(x,\pi^{\star}(x)) + \int u^{\star}(y)Q'(dy \mid x,\pi^{\star}(x)),$$

i.e., for the strategy $\pi^{\star} \in LS_D$ (5.10) is fulfilled and therefore, by Theorem 5.14, $\pi^{\star}$ is a locally optimal strategy. $\qquad\square$

Remark 5.17. Under the conditions and assumptions of Theorem 5.16, we now conclude that in the set of optimal strategies, there is at least one deterministic strategy, which makes decisions on the basis of the actual state only. The optimal value $\phi(x, \pi^{\star})$ under this strategy depends according to the observation of Ross [Ros70] only on the transition kernels $Q(\cdot \mid x, a)$ and $\pi^n(\cdot, x)$ and on mean conditional sojourn times $\tau(x, a)$ and mean conditional rewards $r(x, a)$. Thus $\phi(x, \pi^{\star})$ is insensitive against variations of the shape of the conditional sojourn times distribution and the conditional rewards distribution as long as their conditional means remain invariant. Therefore we can restrict the computation of $\phi(x, \pi^{\star})$ without loss of generality to processes with deterministic sojourn times and special reward functions as given at the end of Section 5.1.3. $\qquad\odot$

Our next theorem provides easy to check smoothness conditions for the data of the semi-Markov models such that the condition 1) and 2) of Theorem 5.16 are fulfilled.

Theorem 5.18. Consider a semi-Markov random field according to Definition 5.9. Recall that the local state spaces X_i and local decision spaces A_i, and therefore the global state space X and the global decision space A as well, are compact metric spaces with countable basis. Let the map A, which associates with any state x the time invariant restriction set $A(x)$, i.e.,

$$A\colon X \longrightarrow 2^A - \{\emptyset\}, \qquad x \to A(x) = \prod_{i \in V} A_i(x),$$

be continuous and let Assumptions 5.11, 5.12, and 5.13 hold. Assume further that

    1) the mean value functions $r(x, a)$ and $\tau(x, a)$ are continuous in $(x, a) \in \bar{\kappa}$;

    2) the transition probability $Q(\cdot \mid x, a)$ is weakly continuous in $(x, a) \in \bar{\kappa}$.

Then there exists an optimal strategy $\pi^\star$ in LS_D and the function $\pi^\star\colon X \to A$ can be selected from Baire class 1. $\odot$

Proof: The proof uses ideas from Theorem 3 in [GS72b] to apply our Theorem 5.16. To do this, we need the following property, which combines [GS72b, Lemma 1, 2]:

    Let A be compact and the mapping $A\colon X \to 2^A - \{\emptyset\}$ continuous. Let $u\colon \bar{\kappa} \subseteq X \times A \to \mathbb{R}$ be bounded and continuous. Then the function $u^*(x) = \sup_{a \in A(x)} u(x, a)$ is continuous.

    Applying the conditions 1) and 2) and this property to the maximization operator $\mathfrak{U}$ in (5.15) yields that $\mathfrak{U}$ maps the space $B(X) \subseteq M(X)$ of real valued bounded continuous functions on X into itself. So condition 1) of Theorem 5.16 is fulfilled. For condition 2) of Theorem 5.16 it suffices to show that for any $u \in C(X)$ the function $A_u\colon X \to 2^A - \{\emptyset\}$, $x \to A_u(x)$ is upper semicontinuous. According to [Kur69, p. 61] it is sufficient to prove that for fixed $u \in C(X)$ the following holds:

    If for $n \to \infty$ we have $(x^n, a^n) \to (x, a)$ with $x^n \in X$ and $a^n \in A_u(x^n)$, then it follows $a \in A_u(x)$.

By definition $a^n \in A_u(x^n)$ means

$$r(x^n, a^n) + \int u(y)\mathsf{Q}'(dy \mid x^n, a^n)$$

$$= \sup_{\tilde{a} \in A(x^n)} \left\{ r(x^n, \tilde{a}) + \int u(y)\mathsf{Q}'(dy \mid x^n, \tilde{a}) \right\}.$$

Now the left side of the equation is a continuous function in (x, a), while the right side is a continuous functions in x. Letting $n \to \infty$ we have

$$r(x, a) + \int u(y)\mathsf{Q}'(dy \mid x, a) = \sup_{\tilde{a} \in A(x)} \left\{ r(x, \tilde{a}) + \int u(y)\mathsf{Q}'(dy \mid x, \tilde{a}) \right\},$$

i.e., $a \in A_u(x)$. From Theorem 5.16 we conclude the existence of an optimal local stationary deterministic strategy $\pi^\star \in LS_D$. $\qquad\square$

5.2 LOCAL CONTROL OF INTERACTING MARKOV JUMP PROCESSES IN CONTINUOUS TIME

In this section we consider Markov jump processes. For the case of denumerable state space and an action space that is a Borel space, recent results for continuously controlled processes are proved in [GHL03]; for further literature, see the references there.

Our focus will be on system properties that are consequences of the graph-structured state and action spaces. In case of state and action spaces that are compact metric spaces with countable basis, the results on semi-Markov processes from Section 5.1 apply if the holding times of the process are memoryless.

5.2.1   Markov jump processes with locally interacting components

For systems with locally interacting coordinates, we introduced in Section 3.1 a formal definition of an interaction structure and a state space that is spatially structured by an interaction graph. As in Section 5.1 we assume now that the local state spaces X_i are compact metric space with countable basis, endowed with Borel-σ-algebra $\mathfrak{X}_i$. The evolution over time of our system with global state space

$$(X, \mathfrak{X}) = \left(\underset{i \in V}{\times} X_i, \sigma \left\{ \underset{i \in V}{\times} \mathfrak{X}_i \right\} \right)$$

is described by a stochastic process $\eta = \left(\eta^t \colon t \in [0, +\infty) \right)$ with random fields as one-dimensional marginals in time. So $\eta^t = \left(\eta_k^t \colon k \in V \right)$ where the subscript k in η_k^t refers to the vertex k, and η_k^t therefore denotes the marginal state at time t and node k of the vector valued process $\eta = \left(\eta^t \colon t \geq 0 \right)$.

The Markov processes we start from are jump processes and therefore special semi-Markov processes in continuous time as described in Section 5.1.1. The important special feature of Markov processes is that the holding times in the respective states of the processes are memoryless, i.e., are exponentially distributed with parameters depending on the actual state. We have in Equation (5.2)

$$
\begin{aligned}
\mathrm{T}(s \mid x, y) &= \Pr \left\{ \tau^n \leq s \mid \xi^n = x, \xi^{n+1} = y \right\} \\
&= \Pr \left\{ \tau^n \leq s \mid \xi^n = x \right\} \\
&= 1 - e^{-\lambda(x)s}, \quad x, y \in X, \; s \geq 0.
\end{aligned}
$$

Definition 5.19 (Markov jump process).   A homogeneous Markov jump process in continuous time is a semi-Markov process $\eta = \left(\eta^t \colon t \geq 0 \right)$ with state space $(X, \mathfrak{X})$ and paths that are right continuous and have left-hand limits (cadlag paths) defined by a sequence $(\xi, \tau) = \left\{ (\xi^n, \tau^n), n = \right.$

$0, 1, \ldots$ }. We assume that the process $\eta = (\eta^t \colon t \geq 0)$ has stationary transition probabilities P:

$$P(t; x, C) := \Pr\left\{\eta^{t+s} \in C \mid \eta^s = x\right\} \quad \forall\, s \in \mathbb{R}_+ = [0; \infty), x \in X, C \in \mathfrak{X}.$$

Note that the right side is independent of s. The X-valued sequence $\{\xi^n, n = 0, 1, \ldots\}$ is the sequence of states the process entered just after the jump instants. The $\mathbb{R}_+$-valued random variables $\{\tau^n, n = 1, 2, \ldots\}$ are the interjump times. If $\{\xi^n = x_n, n = 0, 1, \ldots\}$ is given, then $\{\tau^n, n = 1, 2, \ldots\}$ is a sequence of independent exponentially distributed variables with parameters $\lambda(x_n)$. To be more precise with the construction:

- The sequence of jump times of η is $\sigma = \{\sigma^n \colon n = 0, 1, \ldots\}$, given by $\sigma^0 = 0$, and $\sigma^n = \sum_{i=1}^{n} \tau^i$, $n \in \mathbb{N}$. Then for $t \in [\sigma^n, \sigma^{n+1})$ we have $\eta^t = \xi^n$, $n \in \mathbb{N}$.
- The one-step transition probability of the embedded jump chain is the Markov kernel $(Q(C \mid x) \colon C \in \mathfrak{X}, x \in X)$ of the sequence $\{\xi^n = \eta^{\sigma^n}, n = 0, 1, \ldots\}$.
- The intensity function $\lambda \colon X \to \mathbb{R}_+$ is a bounded measurable function of x with $\lambda(x) < \Lambda < \infty$ for all $x \in X$.

The transition probabilities P of the homogeneous Markov jump process $\eta = (\eta^t \colon t \geq 0)$ fulfill for $x \notin C$

$$P(t; x, C) = \lambda(x) \int_0^t e^{-\lambda(x)s} ds \int_X Q(dy \mid x) P(t - s; y, C) \qquad (5.18)$$

and for $x \in C$

$$P(t; x, C) = e^{-\lambda(x)t} + \lambda(x) \int_0^t e^{-\lambda(x)s} ds \int_X Q(dy \mid x) P(t-s; y, C). \quad (5.19)$$

It follows from (5.18)–(5.19)

$$\frac{\partial P(t; x, C)}{\partial t} = \lambda(x) \left[-P(t; x, C) + \int_X Q(dy \mid x) P(t; y, C) \right].$$

A continuous time homogeneous Markov jump process $\eta = \left(\eta^t : t \in \mathbb{R}_+\right)$ can be defined as Markov process, for which the weak infinitesimal operator is given by

$$[\mathfrak{Q}f(\cdot)](x) = \lambda(x) \left[-f(x) + \int_X Q(dy \mid x) f(y)\right] \qquad (5.20)$$

on his domain, i.e., a subset of $M(X)$, the set of measurable bounded functions on X. ⊙

In the uncontrolled case, our interest focusses on interacting jump processes, which in the Markov setting are similarly defined as in the case of general semi-Markov processes via the embedded jump chains. The following definition merely recalls Definition 5.2 in the Markovian framework.

Definition 5.20 (Synchronized and local transition kernels). A continuous time Markov jump process $\eta = \left(\eta^t : t \geq 0\right)$ with associated process $(\xi, \tau) = \left\{(\xi^n, \tau^n), n = 0, 1, \dots\right\}$ and with state space $(X, \mathfrak{X}) = \left(\underset{i \in V}{\times} X_i, \sigma\left\{\underset{i \in V}{\times} \mathfrak{X}_i\right\}\right)$ is local if (5.3) holds and is synchronous if (5.4) holds. η is then called a *continuous time Markov jump process with locally interacting synchronous components over* (Γ, X), shortly, a *Markov jump random field.* ⊙

Markov jump processes with discrete state space are of special interest to many of the applications we described in the introduction, e.g., the population dynamics of our very first example fits into this class [Ren93], and the stochastic network processes that encompass the population migration processes as well [Kel79]. In the following we therefore describe examples from queueing theory and from the theory of multidimensional birth-death processes where local control is applied. For more details on network theory and generalized birth-death and migration processes, see

[Kel79, Dad01]. It will be seen that queueing network processes, although showing a stringent local behavior with respect to a natural underlying graph structure will not fully meet our property (5.4) and the subsequent (5.24). But the generalized birth death processes completely meat (5.24). We always assume that the describing Markov processes have right continuous paths with left-hand limits (cadlag paths) and that the dynamics of the process are completely specified by the respective Q-Matrices $\mathfrak{Q} = (q(m,n)\colon m,n \in \mathbb{N})$. We start with the description of a service station in isolation, which will determine the local node structure of the networks and of the multidimensional birth-death processes as well.

Definition 5.21 (State dependent single server queue). We consider a single server of $M/M/1/\infty$ type where indistinguishable customers arrive one by one. Customers finding the server free enter service immediately, while customers finding the server busy enter the waiting room, which has an infinite number of waiting places. Unless otherwise specified, the waiting room is organized according to the First-Come-First-Served regime (FCFS): If a service expires the served customer immediately leaves the system and the customer at the head of the queue, if any, enters service and other customers in line are shifted one step ahead. We always assume that these shifts take zero time.

   If there are n customers in the system, then time until the next arrival passes with intensity $\lambda(n)$, and (if $n > 0$) the customer in service is served with intensity $\mu(n)$. Given the number of customers in the system, which we henceforth call the queue length, the actual residual service and residual interarrival times are independent of the past and independent of another.

   We denote by $\eta = (\eta^t\colon t \geq 0)$ the random queue length process of the system.   ⊙

Remark 5.22. From the description it follows that to describe the evolution of the state dependent single server queue by a Markov process, it suffices to record the queue length, i.e., η defined on a suitable probability space $(\Omega, \mathcal{F}, \mathrm{Pr})$ with state space $\mathbb{N}$ is a strong Markov process.

We conclude: The holding time in state n is exponentially distributed with parameter $\lambda(n) + \mu(n)$ and the subsequent jump decision is independent of the holding time, and with probability $\frac{\lambda(n)}{\lambda(n) + \mu(n)}$ the next jump is caused by an arrival to $n+1$, with probability $\frac{\mu(n)}{\lambda(n) + \mu(n)}$ the next jump is caused by a service expiring (departure event) to $n-1$. $\odot$

It turns out that the class of models described in Definition 5.21 coincides with the class of (state dependent) birth-death processes from Definition 2.9.

Theorem 5.23. The queue length process η of the state dependent single server queue is a birth-death process. η is ergodic if and only if

$$G = \sum_{n=0}^{\infty} \prod_{i=1}^{n} \frac{\lambda(i-1)}{\mu(i)} < \infty.$$

If η is ergodic, then its unique steady state and limiting distribution is $\pi = \big(\pi(n) \colon n \in \mathbb{N}\big)$ with

$$\pi(n) = G^{-1} \prod_{i=1}^{n} \frac{\lambda(i-1)}{\mu(i)}, \quad n \in \mathbb{N},$$

where $G < \infty$ is the norming constant. $\odot$

We first discuss the structure of classical queueing networks in the light of the requirements for the local and synchronized behavior of Markov interacting random processes and Markov random fields.

Definition 5.24 (Gordon–Newell network). A Gordon–Newell network is a network of service stations (nodes) numbered $\{1, 2, \ldots, J\} := V$.

Station j is a state dependent single server system with infinite waiting room under FCFS regime (see Definition 5.21 for details of the servicing mechanism). There are $I > 0$ customers cycling according to an irreducible Markov matrix $R = \big(r(i,j) \colon i,j = 1,\ldots,J\big)$.

A customer on leaving node i selects with probability $r(i,j) \geq 0$ to visit node j next, and then enters node j immediately, commencing service if he finds the server free; otherwise, he joins the tail of the queue of node j.

Given the departure node i, the customer's routing decision is made independently of the network's history. Customers arriving at node j request for an amount of work (service time) there which is exponentially distributed with mean 1. All requested service times constitute an independent family of random variables.

Let η_j^t denote the number of customers present at node j at time $t \geq 0$, either waiting or in service (local queue length at node j). Then $\eta^t :=$ $\big(\eta_j^t \colon j = 1,\ldots,J\big)$ is the joint queue length vector of the network at time t. We denote by $\eta = (\eta^t \colon t \geq 0)$ the joint queue length process of the Gordon–Newell network.   ⊙

Theorem 5.25 (see [GN67, Jac63]). We denote by

$$S(I,J) = \big\{(n_1,\ldots,n_J) \in \mathbb{N}^J \colon n_1 + \cdots + n_J = I\big\}$$

the state space of the joint queue length process

$$\eta = \Big(\eta_t \colon (\Omega,\mathcal{F},\mathrm{Pr}) \longrightarrow \big(S(I,J), \mathcal{P}(S(I,J))\big) \colon t \in \mathbb{R}_+\Big)$$

for the Gordon–Newell network defined above.

Then η is a Markov process with Q-matrix $\mathfrak{Q} = \big(q(y,x) \colon y,x \in S(I,J)\big)$ given by:

For $i, j \in V$, and $x = (n_1, \ldots, n_J) \in S(I, J)$

$$q(n_1, \ldots, n_i, \ldots, n_j, \ldots, n_J; n_1, \ldots, n_i - 1, \ldots, n_j + 1, \ldots, n_J)$$
$$= \mu_i(n_i) r(i, j), \quad \text{if} \quad n_i > 0,$$
$$\text{and} \quad q(x, x) = - \sum_{y \in S(I,J) - \{x\}} q(x, y),$$
$$q(x, y) = 0 \quad \text{otherwise.}$$

η is irreducible, conservative, nonexplosive, and ergodic.

Let $\eta = (\eta_1, \ldots, \eta_J)$ denote the unique probability solution of the traffic equation

$$\eta = \eta R.$$

The unique stationary and limiting distribution $\pi = \pi(I, J)$ of η on $S(I, J)$ is

$$\pi(I, J)(n_1, \ldots, n_J) = G(I, J)^{-1} \prod_{j=1}^{J} \left\{ \prod_{k=1}^{n_j} \frac{\eta_j}{\mu_j(k)} \right\},$$
$$(n_1, \ldots, n_J) \in S(I, J),$$

where $G(I, J)$ is the norming constant.   ⊙

Remark 5.26. The definition of the Gordon–Newell network refers only to the set of nodes that represent the nodes of the underlying graph. For the structure graph $\Gamma = (V, B)$ the edges

$$\{i, j\} \in B \iff [r(i, j) > 0 \lor r(j, i) > 0]. \tag{5.21}$$

are defined via the routing matrix R of the customers.   ⊙

Although the network processes act in an obvious sense locally with respect to this neighborhood structure, the transition probabilities of the embedded jump chain are not locally determined and synchronized in the sense of Definition 5.40 and Definition 5.2. This is a result of the rather simple evolution of the network that only allows one service to end at a time instant. But it is direct to generalize the transition behavior in a direction that makes it more in line with the synchronization property we are looking for.

Definition 5.27 (Generalized Gordon–Newell network). A generalized Gordon–Newell network is a network of nodes $\{1, 2, \ldots, J\} := V$ with state dependent single servers from Definition 5.21. $I > 0$ customers cycle according to an irreducible Markov matrix $R = (r(i,j): i,j = 1, \ldots, J)$. (See Definition 5.24.)

Let η_j^t denote the number of customers present at node j at time $t \geq 0$, either waiting or in service (local queue length at node j). Then $\eta^t := (\eta_j^t: j = 1, \ldots, J)$ is the joint queue length vector of the network at time t. We denote by $\eta = (\eta^t: t \geq 0)$ the joint queue length process of the generalized Gordon–Newell network.

The time evolution of the network is as follows:

If $\eta^t = (\eta_j^t: j = 1, \ldots, J) = (n_1, \ldots, n_J)$, then the system stays in this state for a time that is exponentially distributed with parameter $\lambda(n_1, \ldots, n_J) > 0$. If this holding time expires, then at each node there is a decision (for each node, independent of the network's history) whether the ongoing service (if any, i.e., if $n_i > 0$) is finished (with probability $\gamma(i) > 0$) or continues (with probability $1 - \gamma(i) > 0$). If the service at station i expires, then the served customer 'on leaving node i selects with probability $r(i,j) \geq 0$ to visit node j next, and then enters node j immediately, commencing service if he finds the server free; otherwise, he joins the tail of the queue of node j. Given the departure node i, the customer's routing decision is independent of the network's history. $\odot$

Remark 5.28. It follows directly that η is irreducible, conservative, nonexplosive, and ergodic, so a unique stationary and limiting distribution exists.

In a manufacturing context we can interpret the decisions whether service is finished or continued (made at the end of the holding times) to be the activities of the local controllers which reside at the nodes. α_j^n is the decision of the controller at node j at the n-th jump instant of the process and indicates whether the manufacturing process was successful (with probability $\gamma(i) > 0$) or not (with probability $1 - \gamma(i) > 0$).

The transition probability of the embedded jump chain is now local, and from the construction the jump decisions at the nodes are made always at the same time instant in parallel, but it is easily seen that in general the synchronization property which includes conditional independence of the transitions are not completely fulfilled because in the transition probabilities of the jump chain occur the holding time parameters $\lambda(x)$, see (5.4) and (5.24). $\qquad\qquad\qquad\odot$

From the last remark we conclude that we cannot directly apply the result of Corollary 5.41 although to a certain extend the generalized Gordon–Newell network looks like a Markov jump random field from Definition 5.20 with (5.3) and (5.4).

We therefore describe now a supplementary variable technique using an extended state space which makes the time development of the system in a formal sense fulfilling the properties of a Markov jump random field. The main idea is to include for each node (service station) j into the local state description the local queue length n_j there, the decision value whether at the *next* jump instant of the system the ongoing service expires ($= 1$) or is continued for at least one further period ($= 0$), and the direction of the routing (if any) which might occur after a possible departure from j at the *next* jump instant.

Definition 5.29 (Supplemented Gordon–Newell network). A generalized supplemented Gordon–Newell network is a network of state dependent single servers from Definition 5.21 and with nodes $\{1, 2, \ldots, J\} := V$. $I > 0$ customers cycle according to an irreducible Markov matrix $R = (r(i,j) \colon i, j = 1, \ldots, J)$. (See Definition 5.24.) The underlying routing graph is given in (5.21).

Let for time $t \geq 0$ denote by $\eta_j^t = (\nu_j^t, \vartheta_j^t, \varrho_j^t)$ the number ν_j^t of customers present at node j at either waiting or in service (local queue length at node j), the decision ϑ_j^t whether at the next jump instant the customer in service (if any) will depart from j, and the direction $\varrho_j^t \in V$ where this moving customer will arrive then. Then $\eta^t := \left((\nu_j^t, \vartheta_j^t, \varrho_j^t) : j = 1, \dots, J \right)$ is the joint supplemented queue length vector of the network at time t. We denote by $\eta = (\eta^t : t \geq 0)$ the joint supplemented queue length process of the supplemented generalized Gordon–Newell network.

The time evolution of the network is as follows: If

$$\eta^t = \left((\nu_j^t, \vartheta_j^t, \varrho_j^t) : j = 1, \dots, J \right)$$
$$= (n_1, t_1, r_1; \dots; n_i, t_i, r_i; , \dots; n_j, t_j, r_j; \dots; n_J, t_J, r_J),$$

then the system stays in this state for a time that is exponentially distributed with parameter $\lambda(n_1, \dots, n_i, \dots, n_j, \dots, n_J) > 0$. During this holding time, the decision makers at the nodes determine the values $\alpha_k^t \in \{0, 1\}$, $k \in V$.

Independent of the network's history it is decided whether the ongoing service (if any, i.e., if $n_i > 0$) will be finished (with probability $\mathrm{Pr}\left(\alpha_i^t = 1 \right) = \gamma(i) > 0$) or not (with probability $\mathrm{Pr}\left(\alpha_i^t = 0 \right) = 1 - \gamma(i) > 0$) for all $i \in V$.

If this holding time expires, then at each node j in a first step, the queue length is updated according to the values $\left\{ (\eta_i^t, \vartheta_i^t, \varrho_i^t) = (x_i, t_i, r_i) : i \in \widetilde{N}(j) \right\}$. This yields ν_j^{t+1}.

Then for each node j the value ϑ_j^t is updated to $\vartheta_j^{t+1} = \alpha_j^t$.

And finally the value ϱ_j^t is reset to $\varrho_j^{t+1} = k$ with probability $r(j, k) \geq 0$, where given the departure node j the customer's routing decision is made independently of the network's history. $\odot$

Corollary 5.30. The embedded jump chain of the supplemented generalized Gordon–Newell network process η from Definition 5.29 fulfills the locality property (5.3) and the synchronization property (5.4) from Definition 5.20 and so it is a Markov jump random field as defined there.

In addition the jump chain of the controlled network process (η, α) fulfills (5.24) from Definition 5.40 and so (η, α) is a controlled Markov jump random field. $\odot$

Proof: Updating of the queue length according to the values

$$\left\{ (\eta_i^t, \vartheta_i^t, \varrho_i^t) = (x_i, t_i, r_i) \colon i \in \tilde{N}(j) \right\}$$

is purely deterministic at each node, interfering only with neighboring nodes. This is due to the selected state space and the structure of the underlying graph, and the conditioning on η^t. So given η^t, we have conditional independence of the ν_j^{t+1}. The decisions for the α_j^t and therefore the ϑ_j^t are independent of the network's history and independent of another, and the routing decisions are made independent of another as well. This yields the required conditional independence of the synchronization property. $\square$

We start our discussion of multidimensional birth-death processes with a general description of multidimensional Markov processes with state space $\mathbb{N}^J$.

Definition 5.31 (Net migration processes). Consider a Markov process

$$\eta = \left(\eta^t \colon (\Omega, \mathcal{F}, \mathrm{Pr}) \to (X, \mathfrak{X}) \right)$$

with $X = \mathbb{N}^J$ and $\mathfrak{X} = 2^X$, which fulfills the Assumption 2.7. For $x, y \in X$ with $q(x, y) > 0$ let $(y_j - x_j)_+ = b_j$ and $(x_j - y_j)_+ = d_j$ denote the net increment and decrement, of this transition in the j-th coordinate. We call $b = (b_1, \ldots, b_J)$ the net increment vector and $d = (d_1, \ldots, d_J)$ the net decrement vector. Let $F \subseteq \mathbb{N}^J \times \mathbb{N}^J$ denote the set of feasible pairs of joint increment and decrement vectors, i.e., the net transition pairs for η.

Then the transition rates of η, i.e., the entries of the associated Q-matrix $\mathfrak{Q} = \big(q(x,y)\colon x,y \in X\big)$, are of the form

$$q(x,y) = \Lambda(x,b,d),$$
$$\text{for } y = x - d + b, \ x,y \in X, \quad \text{with } q(x,y) > 0, \quad (5.22)$$

and we have $F = \big\{(b,d)\colon \Lambda(x,b,d) > 0 \text{ for some } x \in X\big\}$ for the set of feasible net movements of η. $\qquad\qquad\qquad\qquad\qquad\qquad\qquad\qquad\odot$

Note that defining $\mathfrak{Q}$ from a prescribed function Λ on the set of feasible transition pairs $(x,y) \in X^2$ is not a restriction because x and y determine the net increment and net decrement vector uniquely.

Introducing Description (5.22) for the transition rates is often useful to clarify the movements of customers in networks or the migration of individuals between compartments. Much research effort has been dedicated to finding specific functions Λ, which are well suited for describing real problems and which admit to compute performance measures of the system. The most important singular performance quantity is in almost all cases the steady state distribution; see the result for Gordon–Newell networks above. The existence of a steady state for the describing process guarantees that the system stabilizes over time.

Note that this general class of net migration processes in Definition 5.31 encompasses the network processes described above. We now direct our considerations into another field of processes.

Definition 5.32 (Migration and birth-death process). Let η have transition rates (5.22). Then η is called a generalized migration process.

Let η have transition rates

$$q(x,y) \qquad\qquad\qquad\qquad\qquad\qquad\qquad\qquad\qquad\qquad\qquad (5.23)$$

$$= \begin{cases} \Lambda(x,b,\mathbf{0}) = \beta(x,b), & \text{if } x,y = x+b \in X \text{ for some } b \in B; \\ \Lambda(x,\mathbf{0},d) = \delta(x,d), & \text{if } x,y = x-d \in X \text{ for some } d \in D; \\ 0, & \text{otherwise,} \end{cases}$$

where $\mathbf{0}$ is the $|V|$-dimensional vector with all entries 0, and $B, D \subseteq \mathbb{N}^J$ are the sets of birth vectors, death vectors, respectively, for individuals, and $F = (B \times \{\mathbf{0}\}) \cup (\{\mathbf{0}\} \times D)$. $\beta(\cdot)$ are the birth rates, $\delta(\cdot)$ the death rates. In such processes, births do not occur simultaneously with deaths and vice versa, but births in several coordinates may occur simultaneously, and deaths as well.

A Markov process η with transition rates (5.23) is called generalized birth-death process. $\qquad\qquad\qquad\qquad\qquad\qquad\qquad\qquad\qquad\qquad\qquad$ ⊙

Remark 5.33. Serfozo [Ser92] reserves the term compound migration and compound birth-death process for reversible Markov processes with transition rates (5.22), (5.23). We do not use this convention here because migration processes usually are not reversible in time — see [Kel79], Chapters 2 and 6. It follows that every Markov process with multidimensional state space $(X, \mathfrak{X}) = \left(\underset{i \in V}{\times} X_i, \sigma \left\{ \underset{i \in V}{\times} \mathfrak{X}_i \right\} \right)$ that fulfills the requirements of Assumption 2.7 is a generalized migration process. $\qquad\qquad\qquad\qquad\qquad$ ⊙

We are now in a position to define multidimensional birth-death processes which fulfill the requirements of (5.4) and (5.24).

Proposition 5.34. Let $\Gamma = (V, B)$ denote an undirected finite graph without loops and double edges and let

$$\eta = \left(\eta^t \colon (\Omega, \mathcal{F}, \mathrm{Pr}) \to (X, \mathfrak{X}) = \left(\mathbb{N}^V, 2^{\mathbb{N}^V} \right) \right)$$

be a Markov process with Q-matrix $\mathfrak{Q} = \left(q(x, y) \colon x, y \in X \right)$. Assume that η is a multidimensional birth-death process with $\mathfrak{Q}$ given in the definition via net increments and decrements in (5.23) from Definition 5.32 as follows: The set of possible net births and deaths vectors are of product structure

$$B = \underset{i \in V}{\times} B_i, \quad D = \underset{i \in V}{\times} D_i,$$

and for all $x \in X$ and all $j \in V$ there exist probability measures

$$\beta_j(x_{\widetilde{N}(j)}; \cdot) \text{ on } B_j \quad \text{and} \quad \delta_j(x_{\widetilde{N}(j)}; \cdot) \text{ on } D_j$$

and a bounded function $\lambda \colon X \to (0, \infty)$ such that

$q(x, y)$

$$= \begin{cases} \Lambda(x, b, \mathbf{0}) = \beta(x, b) = \prod\limits_{j \in V} \beta_j \left(x_{\widetilde{N}(j)}; b_j \right) \cdot \lambda(x), \\ \quad \text{if} \quad x, y = x + b \in X \quad \text{for some} \quad b = \left(b_j \colon j \in V \right) \in B; \\ \Lambda(x, \mathbf{0}, d) = \delta(x, d) = \prod\limits_{j \in V} \delta_j \left(x_{\widetilde{N}(j)}; d_j \right) \cdot \lambda(x), \\ \quad \text{if} \quad x, y = x - d \in X \quad \text{for some} \quad d = \left(d_j \colon j \in V \right) \in D; \\ 0, \quad \text{otherwise.} \end{cases}$$

Then the embedded jump chain of the multidimensional birth-death process η fulfills the locality and synchronization properties (5.3) and (5.4) from Definition 5.20 and so it is a Markov jump random field as defined there. $\odot$

In Section 3.2 we discussed the reversibility of discrete time Markov chains in connection with theorems that provided conditions that guarantee that local and synchronous processes have Markov fields as one-dimensional marginals in time; see Theorems 3.14 and 3.15. The same question naturally arises in the context of generalized migration processes and subclasses thereof. We describe an interesting result from [Kel79, Section 9.2], which we present in the notation of multidimensional birth-death processes as in Proposition 5.34.

Definition 5.35 (Spatial process). Let $\Gamma = (V, B)$ denote an undirected finite graph without loops and double edges and let

$$\eta = \left(\eta^t \colon (\Omega, \mathcal{F}, \mathrm{Pr}) \to (X, \mathfrak{X}) = \left(\mathbb{N}^V, 2^{\mathbb{N}^V} \right) \right)$$

be a Markov process with Q-matrix $\mathfrak{Q} = \big(q(x,y)\colon x, y \in X\big)$. Assume that η is a multidimensional birth-death process with $\mathfrak{Q}$ given in the definition via net increments and decrements in (5.23) from Definition 5.32 as follows: The set of possible net births and deaths vectors are of product structure

$$B = \underset{i \in V}{\times} B_i, \quad D = \underset{i \in V}{\times} D_i,$$

and for all $x \in X$ and all $j \in V$ there exist probability measures

$$\beta_j(x_{\widetilde{N}(j)}; \cdot) \text{ on } B_j \quad \text{and} \quad \delta_j(x_{\widetilde{N}(j)}; \cdot) \text{ on } D_j$$

and a bounded function $\lambda\colon X \to (0, \infty)$ such that

$$q(x,y)$$
$$= \begin{cases} \Lambda(x, b, \mathbf{0}) = \beta(x, b) = \beta_j\left(x_{\widetilde{N}(j)}; b_j\right) \cdot \lambda(x), \\ \qquad \text{if } x, y = x + b \in X \text{ for some } b = (b_i\colon i \in V) \in B, \\ \qquad\qquad\qquad \text{with } b_i = 0, \ \forall \, i \neq j; \\ \Lambda(x, \mathbf{0}, d) = \delta(x, d) = \delta_j\left(x_{\widetilde{N}(j)}; d_j\right) \cdot \lambda(x), \\ \qquad \text{if } x, y = x - d \in X \text{ for some } d = (d_i\colon i \in V) \in D, \\ \qquad\qquad\qquad \text{with } d_i = 0, \ \forall \, i \neq j; \\ 0, \qquad \text{otherwise.} \end{cases}$$

We assume further that any change of the state in one coordinate, say j, can be realized by transitions that do not alter the local states at nodes other than j. Then η is a spatial process. $\odot$

Remark 5.36. A spatial process is not, in general, a Markov jump random field according to Definition 5.20. $\odot$

Theorem 5.37 (see [Kel79, Theorem 9.3]). The equilibrium distribution of a reversible spatial process is a Markov field. $\odot$

5.2.2 Markov jump processes with local and synchronous transition mechanisms and control at jump instants

In this section we introduce a controlled Markov process with locally interacting synchronous components under step control. See Section 2.3.1 for the case of unstructured state space, or [GS69] and the more recent [GHL03]. Note that the step control mechanism described in Section 2.3.1 allows *new decisions* only at the jump times $(\sigma^n \colon n \in \mathbb{N})$ of the controlled process.

If a continuous time Markov jump process η and its associated process (ξ, τ) (Definition 5.19) from their very construction are related to the neighborhood system $\{N(k) \colon k \in V\}$ of (V, B), it is natural to assume that with respect to the admissible control, a similar restriction of availability of information is in force. To formalize this, we introduce the Markov property in space for the jump kernel of the continuous time Markov process along the lines of Definition 3.10 and for the decision making as well; see Definition 3.18 for locality of strategies. We start from the general semi-Markov process setting and consider controls that are step controls as described in Section 2.3 and action spaces with local restrictions, which are described in Definition 5.3. Recall the following notation.

Definition 5.38 (Action spaces and local restrictions). The sequence of decision instants is $\sigma = (\sigma^n, n = 0, 1, \dots)$, the jump times of $\eta = (\eta^t \colon t \geq 0)$. (See Remark 5.5 for a comment on prescribing this set of feasible decision instants.)

(1) The set of actions (control values) usable at control instants is a measurable space $(A, \mathfrak{A}) = \left(\underset{i \in V}{\times} A_i, \underset{i \in V}{\bigotimes} \mathfrak{A}_i \right)$ over Γ, where A_i is a set of possible actions (decisions) for vertex i. We assume that A_i is a compact metric space with countable basis.

(2) If for the decision maker at node i at time σ^n under state $\xi^n = x$ the set of control actions is restricted to $A_i^n(x) \subset A_i$, we call $A_i^n(x)$ the set of admissible actions (decisions) at time σ^n in state x. We always assume that the restriction sets are time invariant and therefore depend on the actual state of the system only. We denote $A_i(x) := A_i^n(x) \subset A_i$, for all decision instants σ^n. $\odot$

Definition 5.39 (Local strategies). Let α_i^t denote the action chosen by the decision maker at node i at the instant t, $\alpha^t := \left(\alpha_i^t : i \in V\right)$ the joint decision vector at time t, $t \in \mathbb{R}_+$. We assume that the strategies are prescribed according to Definition 5.7 **(2)** with probability measures $\pi_i^n(\cdot \mid \cdot, \ldots, \cdot)$ and that the decision makers act conditionally independent of another as described in Definition 5.7 **(3)**.

The class of all these admissible local (randomized) strategies is denoted by LS. For the further classification of local strategies, see Definition 5.8, which applies here as well. $\odot$

The law of motion of the system is characterized by a set of time invariant transition kernels. Whenever the state of the system is $\eta^t = x^t$ and decision $\alpha^t = a^t$ is made, the transition kernel is $\Pr\{\eta^{t+s} \in C \mid \eta^t = x^t, \alpha^t = a^t\} =: \mathsf{P}\left(x^t, s, C; a^t\right)$, which is homogeneous in time. Then, given $\left\{\eta^t = x^t, \alpha^t = a^t\right\}$ the state of the system $\eta^{t+s} = x^{t+s}$ in time s is sampled according to $\mathsf{P}\left(x^t, s, \cdot; a^t\right)$.

When applying a control policy π to a continuous time Markov jump process with interacting components $\eta = (\xi, \tau)$ (Markov jump random field), we call the pair (η, π) a controlled version of η, using strategy π. If we allow only Markovian strategies in our control, the controlled process will be Markovian as well. The following definition summarizes the discussion and recalls merely the Definition 5.9, with suitable adaption to the present framework.

Definition 5.40. A pair (η, π), respectively a triple (ξ, τ, π), is called a controlled Markov jump process with locally interacting synchronous components with respect to the finite interaction graph $\Gamma = (V, B)$, if the following holds:

- $\xi = (\xi^n \colon n \in \mathbb{N})$ is a stochastic process with state space $X = \underset{i \in V}{\times} X_i$,

$\tau = (\tau^n \colon n \in \mathbb{N})$ is stochastic process with state space $\mathbb{R}_+$;

- $\eta = (\eta^t \colon t \geq 0)$ is a stochastic process with state space $X = \underset{i \in V}{\times} X_i$,

and these processes are connected pathwise as follows:

The $\mathbb{R}_+$-valued random variables τ^n are the interjump times distributed exponential with parameter $\lambda(x, a)$, which measurably depends on (x, a), i.e., on the system state $\eta^t = x \in X$ at jump instant t if decision $\alpha^t = a$ is in force, and the sequence $\{\xi^n, n = 0, 1, \dots\}$ is the sequence of states of the process entered just after the jump instants.

- $\sigma = \{\sigma^n \colon n = 0, 1, \dots\}$ with $\sigma^0 = 0$, and $\sigma^n = \sum_{i=1}^n \tau^i$, $n \in \mathbb{N}$, is the increasing sequence of jump times. Then if $t \in [\sigma^n, \sigma^{n+1})$ we have $\eta^t = \xi^n$, $n \in \mathbb{N}$.

- $\lambda(x, a)$ is a non-negative and bounded: $0 < \lambda < \lambda(x, a) < \Lambda < \infty$ for all $x \in X$, $a \in A$.

We assume that $\pi = (\pi_i \colon i \in V)$ is an admissible local pure Markov strategy, and that the conditional distribution function of η^t is $\mathrm{P}(x, s, \cdot; a)$, which is Borel measurable on $X \times \mathbb{R}_+$, and that the transition probabilities of ξ fulfill

$$
\begin{aligned}
\Pr\{\xi_K^{n+1} &\in C_K \mid \xi^n = x, \alpha^n = a\} \\
&= \prod_{j \in K} \Pr\{\xi_j^{n+1} \in C_j \mid \xi^n = x, \alpha^n = a\} \\
&= \prod_{j \in K} \Pr\left\{\xi_j^{n+1} \in C_j \mid \xi_{\widetilde{N}(j)}^n = x_{\widetilde{N}(j)}, \alpha_j^n = a_j\right\} \\
&= \prod_{j \in K} \mathsf{Q}_j\left(C_j \mid x_{\widetilde{N}(j)}, a_j\right) \\
&= \mathsf{Q}_K(C_K \mid x, a), \qquad K \subseteq V, \ y \in X, \ a_j \in A_j\left(x_{\widetilde{N}(j)}\right). \quad (5.24)
\end{aligned}
$$

The transition probabilities $Q_K(C_V \mid x, a)$ are the local jump probabilities (which are not, in general, Markov). If $K = V$, we shall write $Q_V(C_V \mid x, a) = Q(C \mid x, a)$.

The Markov kernel $Q = \prod_{j \in V} Q_j$ is said to be local and synchronous. For comments on the equations in (5.24), see the comments in the general semi-Markov setting on page 164 and on the standard Markov chain setting for interacting processes after Definition 3.21. ⊙

Applying now the results of Section 5.1.4 to the Markovian setting, we obtain under the average reward optimality criterion from Definition 5.10:

Corollary 5.41. If the smoothness conditions of Theorem 5.18 hold, then in the class LS of local strategies there exist a stationary deterministic optimal local policy. ⊙

5.2.3 Markov jump processes with local and synchronous transition mechanisms and deterministic control

In this section we restart from a Markov jump process $\eta = (\eta, \tau)$ with synchronized and localized transition kernels (Definition 5.19, 5.20) and introduce a different control mechanism. While in Section 5.2.2 the applied controls could only changed at jump instants of the process, we now prescribe a (then fixed) deterministic sequence of time points when the control is renewed. With respect to selecting the control points we work with an open-loop control principle, while the selection of the decision is according to closed-loop principles. These controls are easier to implement and therefore are of special interest. We follow in the main presentation the ideas that are developed in [Mil68] for the case of finite state space without a localized graph structured space.

Let a time homogeneous Markov jump process $\eta = (\eta^t : t \geq 0)$ be given as in Definition 5.19 with state space

$$(X, \mathfrak{X}) = \left(\underset{i \in V}{\times} X_i, \sigma \left\{ \underset{i \in V}{\times} \mathfrak{X}_i \right\} \right).$$

We assume that the X_i and so X are compact metric spaces with countable basis. We further assume that the process is local, i.e., (5.3) holds and synchronous, i.e., (5.4) holds. So η is from its very definition a *continuous time Markov jump process with locally interacting synchronous components over* (Γ, X), shortly, a *Markov jump random field*.

In this section we consider only finite action spaces and controls using action spaces with local restrictions, which are described in Definition 5.3. Recall the following notation.

Definition 5.42 (Action spaces and local restrictions). Decision instants are times $t \in [0, \infty)$.

(1) The set of actions (control values) usable at control instants is a measurable space $(A, \mathfrak{A}) = \left(\underset{i \in V}{\times} A_i, \underset{i \in V}{\bigotimes} \mathfrak{A}_i \right)$ over Γ, where A_i is a set of possible actions (decisions) for vertex i. We assume that $A_i = \{a_{i,1}, a_{i,2}, \ldots, a_{i,m_i}\}$ is finite and $\mathfrak{A}_i = 2^{A_i}$.

(2) If for the decision maker at node i at time t under state $\eta^t = x$ the set of control actions is restricted to $A_i^t(x) \subset A_i$, we call $A_i^t(x)$ the set of admissible actions (decisions) at time t in state x. We always assume that the restriction sets are time invariant and therefore depend on the actual state of the system only. We denote $A_i(x) := A_i^t(x) \subset A_i$, for all decision instants t. $\odot$

The strategies under consideration are completely deterministic (pure strategies).

Definition 5.43 (Local pure strategies). Let α_i^t denote the action selected by the decision maker at node i at the instant t, $\alpha^t := (\alpha_i^t : i \in V)$ the joint decision vector at time t, $t \in \mathbb{R}_+$. We assume that strategies are prescribed by:

(1a) A strictly increasing sequence $0 = t^0 < t^1 < t^2 < \dots$ of global decision instants, which may be finite or infinite, and if $0 = t^0 < t^1 < t^2 < \dots < t^s$ is a finite sequence, then $t^s = \infty$ is allowed, and

(1b) a sequence of (deterministic) measurable functions $R^1, R^2, \dots$, where $R^n \colon X \to A$.

(2) An admissible local pure Markov strategy $\pi = (\pi^t : t \geq 0)$ is a family of functions $\{(\pi^t \colon X \to A) : t \geq 0\}$ such that $\pi \colon (\mathbb{R}_+ \times X, \mathbb{B}_+ \otimes \mathcal{X}) \to (A, \mathfrak{A})$ is measurable. The coordinate functions of the control $\{\pi^t = (\pi_i^t : i \in V) : t \geq 0\}$ determine the local decisions at the vertices $i \in V$. We assume that the local decisions at i are made on the basis of the full neighborhood $\widetilde{N}(i)$ only. So we have $\pi_i^t(x) = \pi_i^t\left(x_{\widetilde{N}(i)}\right)$.

These policies constitute in the present framework the class LS_{PM}; see Remarks 3.19 and 3.20.

(3) We denote the class of such admissible local pure Markov strategies with deterministic decision times by LS_{PM}^d.

Strategies in $\pi \in LS_{PM}^d$ are encoded as follows. The local policy π_i at node i is determined by a sequence

$$\pi_i = (R_i^1, t^1; R_i^2, t_2; \dots),$$

where $0 = t^0 < t^1 < t^2 < \dots$ is the prescribed sequence of decision instants and R_i^n is the deterministic measurable function on X with values in A_i that yield the decision value that is in force during the time interval $[t^{n-1}, t^n)$ (for the decision maker at node i if the state of the system is x).

We have then $\pi_i^t(x) := R_i^n(x) = R_i^n\left(x_{\widetilde{N}(i)}\right)$ if $t \in [t^{n-1}, t^n)$.

The global decision values are sequences $\pi = (R^1, t^1; R^2, t_2; \dots)$ with $R^n = (R_i^n : i \in V)$.

(4) We use the following concatenation rule for strategies. If

$$\pi = \left(R^1, t^1; R^2, t_2; \ldots \right)$$

is a given strategy, then

$$\pi' = \left(\widetilde{R}^1, t^1; \widetilde{R}^2, t_2; \ldots; \widetilde{R}^j, t^j; \pi \right)$$

is the strategy that applies for $t < t^j$

$$\pi'(t) = \widetilde{R}^n, \quad \text{if } t \in [t^{n-1}, t^n), \quad \text{for } n \in \{1, \ldots, j\}$$

and for $t \geq t^j$

$$\pi'(t) = \pi(t - t^j).$$

(5) We denote by LS_D^d the class of admissible local stationary pure (Markov) strategies.

(6) An admissible pure stationary policy from LS_D^d is encoded as $\left(R^1, t^1 \right) = (R, \infty)$.

We therefore identify such strategy with the function $R \colon X \to A$. $\odot$

We obtain a controlled Markov jump process (η, π) with interacting coordinates from η by incorporating the control $\pi \in LS_{PM}^d$ into the time development of the process. It turns out, that due to the fixed times when the control changes, the resulting process is Markovian but no longer time homogeneous.

If the strategy applied is from LS_D^d, determined by the function $R \colon X \to A$, we obtain a homogeneous Markov jump process with locally interacting coordinates for which weak infinitesimal operator on the space $M(X)$ is given by

$$[\mathfrak{Q}(R)f(\cdot)](x) = \lambda\big(x, R(x)\big)\left[-f(x) + \int f(y) \mathsf{Q}\big(dy \mid x, R(x)\big)\right]. \tag{5.25}$$

If $\pi = \left(R^1, t^1; R^2, t_2; \ldots \right)$ we obtain on any time interval $[t^{n-1}, t^n)$ by the standard construction a time homogeneous Markov jump process

$$(\eta_n, \pi_n) = \Big(\big(\eta_n(t), R^n\big) \colon t \in [t^{n-1}, t^n) \Big)$$

with jump kernel

$$Q_n\big(dy \mid x, R^n(x)\big) = Q\big(dy \mid x, R^n(x)\big)$$

and intensity function $\lambda\big(x, R^n(x)\big)$, which are substituted into (5.20) and (5.25) to obtain the stepwise the infinitesimal generators with domain $M(X)$ on successive time intervals:

$$\big[\mathfrak{Q}(R^n)f(\cdot)\big](x) = \lambda\big(x, R^n(x)\big)\bigg[- f(x) + \int f(y)Q\big(dy \mid x, R^n(x)\big)\bigg].$$

We can always select a version of the (η_n, π_n) process with cadlag paths on $[t^{n-1}, t^n)$. Note that each of these processes has overall bounded intensity functions.

These processes have to be pasted together in a final construction step. See [Mil68] and [Dyn63, Vol. 1, p. 29] for details on the existence and the construction.

5.2.4   Criteria for optimality

Consider a time dependent controlled Markov jump random field (η, α) as defined in Section 5.2.2 with random step control $\pi \in LS$ or in Section 5.2.3 with deterministic control $\pi \in LS_{PM}^d$. If η during $[t, t + \Delta t)$ is in state $x \in X$ and the joint decision $a \in A$ is in force at that time then a reward $r(x, a)\Delta t$ is obtained. $r(x, a)$ is assumed to be a measurable function of (x, a) and we assume for all $x \in X$, $a \in A$

$$r(x, a) < l < \infty.$$

We denote by $\Pr_{x^0}^\pi$ the probability measure that governs the controlled process if strategy π is applied and the process η is started with $\eta^0 = x^0$. With $\mathsf{E}_{x^0}^\pi$ we denote expectations under $\Pr_{x^0}^\pi$.

We consider the following measures to assess the applied strategies:

1. *Discounted total reward:*

$$\psi_\beta\big(x^0, \pi\big) = \mathsf{E}_{x^0}^\pi \int_0^\infty \mathrm{e}^{-\beta t} r\Big(\eta(t), \pi^t\big(\eta(t)\big)\Big) dt,$$

where $\mathrm{e}^{-\beta t}$ is the discount factor, $0 < \beta < \infty$.

This criterion is sometimes called ψ-criterion.

2. *Average reward* (Maximin criterion, see (2.7)):

$$\phi\big(x^0, \pi\big) = \liminf_{T \to \infty} \frac{1}{T} \mathsf{E}_{x^0}^\pi \int_0^T r\Big(\eta(t), \pi^t\big(\eta(t)\big)\Big) dt.$$

This criterion is sometimes called ϕ-criterion.

Another interpretation of the discount factor $\mathrm{e}^{-\beta t}$ is as follows:

The process terminates at a random time that is exponentially distributed with parameter β and independent of the process and the control.

In the next section we investigate controlled Markov jump processes under completely deterministic Markov control in LS_{PM}^d and search for optimal policies in this class. Therefore we formulate the following criteria for policies in LS_{PM}^d. For other classes, the criteria read similarly.

Definition 5.44. (1) For a given discount factor $\beta \in (0, \infty)$ a strategy $\pi^\star \in LS_{PM}^d$ is optimal in the class LS_{PM}^d of admissible local pure Markov strategies with respect to the discounted total reward criterion if

$$\psi_\beta\big(x^0, \pi^\star\big) = \sup_{\pi \in LS_{PM}^d} \psi_\beta\big(x^0, \pi\big), \qquad \forall\, x^0 \in X.$$

Such a $\pi^\star$ is called optimal with respect to the discounted total reward criterion (under β) or ψ-optimal.

(2) A strategy $\pi^\star \in LS_{PM}^d$ is called optimal in the class LS_{PM}^d of admissible local pure Markov strategies with respect to the average reward criterion if

$$\phi\big(x^0, \pi^\star\big) = \sup_{\pi \in LS_{PM}^d} \phi\big(x^0, \pi\big), \qquad \forall\, x^0 \in X.$$

Such a π^* is called optimal with respect to the average reward criterion or ϕ-optimal. ⊙

In the following, we shall in the context of discounted reward problems refer simply to discounted total reward optimality or ϕ-optimality if the selected β is given from the context and fixed.

5.2.5 Existence of discounted-optimal Markov strategies in the class of local deterministic strategies

We fix in the sequel a discount factor $\beta \in (0, \infty)$ that is in force unless otherwise stated. Optimality or the term ψ-optimality refers to that β.

Let $\{T^t(R), t \geq 0\}$ be a semigroup of operators on $M(X)$, the space of real valued measurable bounded functions on $(X, \mathfrak{X})$ corresponding to the homogeneous Markov jump process controlled by strategy $R \in LS_D^d$:

$$\left[T^t(R)f(\cdot)\right](x) = \mathsf{E}_x^R f\big(\eta(t)\big), \qquad f \in M(X).$$

Applying Fubini's theorem, the functional $\psi_\alpha(R; x)$ can be written as

$$\psi_\beta(x, R) = \int_0^\infty e^{-\beta t} \Big[T^t(R)r\big(\cdot, R(\cdot)\big)\Big](x)\, dt.$$

We introduce an operator $L(R; t)$ operating on the space $M(X)$ by

$$\left[L(R; t)f(\cdot)\right](x)$$
$$= \int_0^t e^{-\beta s}\Big[T^s(R)r\big(\cdot, R(\cdot)\big)\Big](x)\, ds + e^{-\beta t}\Big[T^t(R)f(\cdot)\Big](x). \quad (5.26)$$

Recall the concatenation rule for strategies from Definition 5.43 **(4)**. Then for $R \in LS_D^d$, $\pi \in LS_{PM}^d$, and $t > 0$ we have

$$\psi_\beta\big(x, (R, t, \pi)\big) = \Big[L\big(R, t\big)\psi_\beta(\cdot, \pi)\Big](x)$$

and for $\pi' = \big(\widetilde{R}^1, t^1; \widetilde{R}^2, t_2; \ldots; \widetilde{R}^n, t^n\big) \in LS^d_{PM}$ with $t^n < \infty$ and $\pi \in LS^d_{PM}$ we have

$$\psi_\beta\big(x, (\pi', \pi)\big) = \psi_\beta\Big(x, \big(\widetilde{R}^1, t^1; \widetilde{R}^2, t_2; \ldots; \widetilde{R}^n, t^n, \pi\big)\Big) =$$
$$= \Big[L(R^1, t^1) L(R^2, t^2 - t^1) \cdots L(R^n, t^n - t^{n-1}) \psi_\alpha(\pi; \cdot)\Big](x).$$

The operators $L(R, t)$ are monotone, i.e., if $f^1(x) \geq f^2(x)$ for all $x \in X$, then $\big[L(R, t) f^1(\cdot)\big](x) \geq \big[L(R, t) f^2(\cdot)\big](x)$. This follows from the monotonicity of the operators $T^t(R)$, $t \geq 0$.

The proofs of the following theorems are similar to the proofs in Theorems 1 and 2 in [Mil68] for standard controlled Markov chains with finite state and action space.

Theorem 5.45. If for $\pi \in LS^d_{PM}$

$$\psi_\beta(x, \pi) \geq \psi_\beta\big(x, (R, t, \pi)\big)$$

for all $x \in X$, $R \in LS^d_D$, and $t > 0$, then the strategy $\pi \in LS^d_{PM}$ is ψ-optimal. ⊙

Theorem 5.46. If for strategies $R \in LS^d_D$, and $\pi \in LS^d_{PM}$, and some $t \in (0, \infty)$

$$\psi_\beta\big(x, (R, t, \pi)\big) \geq \psi_\beta(x, \pi)$$

holds for all $x \in X$, then

$$\psi_\beta(x, R) \geq \psi_\beta(x, \pi), \qquad x \in X. \qquad ⊙$$

Theorem 5.47. Let

$$D^\pi_R(x) = r\big(x, R(x)\big) + \big[\mathfrak{Q}(R) \psi_\beta(\pi; \cdot)\big](x) - \beta \psi_\beta(x, \pi),$$

where $R \in LS^d_D$, $\pi \in LS^d_{PM}$, $x \in X$. Then for every $t > 0$

$$D^\pi_R(x) \genfrac{(}{)}{0pt}{}{\geq}{\leq} 0 \Rightarrow \big[L(R, t) \psi_\beta(\cdot, \pi)\big](x) \genfrac{(}{)}{0pt}{}{\geq}{\leq} \psi_\beta(x, \pi), \qquad x \in X. \qquad ⊙$$

Proof: From equality (5.26) we have

$$\frac{d}{dt}\big[L(R,t)\psi_\beta(\cdot,\pi)\big](x)$$

$$= \mathrm{e}^{-\beta t}\Big[T^t(R)\Big(r\big(\cdot,R(\cdot)\big) + \mathfrak{Q}(R)\psi_\beta(\cdot.\pi) - \beta\psi_\beta(\cdot,\pi)\Big)\Big](x)$$

$$= \mathrm{e}^{-\beta t}\big[T^t(R)D_R^\pi(\cdot)\big](x).$$

From this and from the relation $\big[L(R,0)\psi_\beta(\cdot,\pi)\big](x) = \psi_\beta(x,\pi)$ follows the assertion of the theorem.   □

Let $R \in LS_D^d$. Denote

$$G(R,x) = G_\beta(R,x) = \Big\{a\colon r(x,a) + \big[\mathfrak{Q}(a)\psi_\beta(\cdot,R)\big](x) > \beta\psi_\beta(x,R)\Big\},$$

where

$$\big[\mathfrak{Q}(a)f(\cdot)\big](x) = \lambda(x,a)\Big[-f(x) + \int f(y)Q(dy \mid x,a)\Big].$$

Let

$$G(R) = G_\beta(R) = \big\{x\colon G_\beta(R,x) \neq \emptyset\big\}.$$

Construct on $G(R)$ a function $a(\cdot)$ with values in A in the following way: $a(x)$ is the element from $A(x)$ with the least index such that

$$r\big(x,a(x)\big) + \big[\mathfrak{Q}(a)\psi_\beta(\cdot,R)\big](x) = \max_{a\in A(x)}\Big\{r(x,a) + \big[\mathfrak{Q}(a)\psi_\beta(\cdot,R)\big](x)\Big\}.$$

It is easy to see that the set $G(R)$ is measurable and the function $a(\cdot)$ is measurable mapping of $G(R)$ into $A(x)$.

Recall that the function $a(\cdot)$ in general depends on β.

Theorem 5.48. Let $R \in LS_D^d$. If $G(R)$ is empty, then strategy R is ψ-optimal. If $G(R)$ is not empty, then strategy

$$R^1(x) = \begin{cases} R(x), & x \notin G(R); \\ a(x), & x \in G(R) \end{cases}$$

is better strategy R in such a way that

$$\psi_\beta(x, R^1) \geq \psi_\beta(x, R), \qquad x \in X. \qquad \odot$$

Proof: Let $G(R)$ be empty. Then for every $x \in X$ and every strategy $R' \in LS_D^d$

$$D_{R'}^R(x) \leq 0$$

holds, and from Theorem 5.47 we have that strategy R is ψ-optimal.

Let $G(R) \neq \emptyset$. Then it is easy to see that on set $X \setminus G(R)$

$$D_{R^1}^R(x) = 0$$

holds, and on $G(R)$

$$D_{R^1}^R(x) > 0$$

holds. From this, due to Theorems 5.47 and 5.46, we finally obtain

$$\psi_\beta(x, R^1) \geq \psi_\beta(x, R), \qquad x \in X. \qquad \square$$

Corollary 5.49. If for strategy R holds

$$\psi_\beta(R; x) = \sup_{R' \in LS_D^d} \psi_\beta(R'; x), \qquad x \in X,$$

then R is ψ-optimal in LS_{PM}^d (for this β). $\qquad \odot$

Corollary 5.49 reduces an optimization problem in the class of local step strategies with predetermined deterministic decision instants and, therefore, of nonhomogeneous controlled Markov processes to an optimization problem in the class of stationary strategies with predetermined deterministic decision instants and, therefore, of homogeneous controlled Markov processes. In the following we shall use the below lemma.

Lemma 5.50. The functional $\psi_\beta(x, R)$ is a continuous function of $R \in LS_D^d$, i.e., if $R^n \to R$ converge pointwise, then

$$\psi_\beta(x, R^n) \to \psi_\beta(x, R)$$

converge in every point $x \in X$.   ⊙

Proof: From boundedness of the operator $\mathfrak{Q}(R)$ on $M(X)$ we have (see [Sko90])

$$T^t(R) = e^{t\mathfrak{Q}(R)} = \sum_{k=0}^{\infty} \frac{t^k}{k!} \mathfrak{Q}^k(R).$$

From $r(y, a) < l < \infty$, $\forall\, (y, a)$, we have by dominated convergence for any x

$$\left[\mathfrak{Q}(R^n) r\big(\cdot, R^n(\cdot)\big)\right](x) \to \left[\mathfrak{Q}(R) r\big(\cdot, R(\cdot)\big)\right](x), \quad n \to \infty.$$

From this we obtain for any $x \in X$, $t \geq 0$.

$$\left[e^{t\mathfrak{Q}(R^n)} r\big(\cdot, R^n(\cdot)\big)\right](x) \to \left[e^{t\mathfrak{Q}(R)} r\big(\cdot, R(\cdot)\big)\right](x), \quad n \to \infty.$$

Computing $\psi_\beta(x, R^n)$ needs another application of dominated convergence.   □

Theorem 5.51. For every fixed $\beta > 0$ there exists in class LS_D^d a ψ-optimal strategy R_β.   ⊙

Proof: According to Corollary 5.49 it suffices to prove the existence of a strategy $R_\beta \in LS_D^d$ such that

$$\psi_\beta(x, R_\beta) = \sup_{R' \in LS_D^d} \psi_\beta(x, R'), \qquad x \in X.$$

Let R be an arbitrary strategy from LS_D^d. If $G(R) = \emptyset$, then R is ψ-optimal (Theorem 5.48) and the statement in this case is proved. If $G(R) \neq \emptyset$, then we can construct a strategy R^1 such that

$$\psi_\beta(x, R^1) \geq \psi_\beta(x, R), \qquad x \in X.$$

Note that the strategy R^1 can be constructed such that

$$D_{R^1}^R(x) = r(x, R^1(x)) + \left[\mathfrak{Q}(R^1)\psi_\beta(\cdot, R)\right](x) - \beta\psi_\beta(x, R)$$
$$= \sup_{R' \in LS_{PM}^d} D_{R'}^R(x), \qquad x \in X.$$

These relations show that the strategy R^1 is a locally maximal improvement of the strategy R in case if R is not optimal.

We apply similar arguments to strategy R^1 and continue iteratively. Thus, we either find during a finite number of steps an optimal strategy or construct a sequence $(R, R^1, R^2, \ldots, R^n, \ldots)$ (with $R^n \in LS_D^d$) of strategies (which depend on β), each of which is better than the previous one. Since LS_D^d is compact in topology of pointwise convergence, we can select from the sequence $\{R^n\}$ a subsequence $\{R^{n_k}\}$ converging to some strategy $R_\beta \in LS_D^d$.

We prove that the strategy R_β can not be improved to some other stationary strategy, and assume the contrary: Then $G(R_\beta) \neq \emptyset$ holds, and, consequently, we find a point $x^0 \in X$, for which the set $G(R_\beta, x^0)$ (see proof of Theorem 5.47) is nonempty. Select some element $a^0 \in G(R_\beta, x^0)$. We have

$$r(x^0, a^0) + \left[\mathfrak{Q}(a^0)\psi_\beta(\cdot, R_\beta)\right](x^0) > \beta\psi_\beta(x^0, R_\beta),$$

i.e.,

$$r\left(x^0, a^0\right) + \left[\mathfrak{Q}\left(a^0\right)\psi_\beta\left(\cdot, R_\beta\right)\right]\left(x^0\right) - \beta\psi_\beta\left(x^0, R_\beta\right) = \Delta\left(x^0, a^0\right) > 0.$$

Due to Lemma 5.50 and Lebesgue's dominated convergence theorem, we conclude that for every $\varepsilon > 0$ we find $K = K\left(x^0, a^0, \varepsilon\right) \in \mathbb{N}$ such that for $k > K$

$$r\left(x^0, a^0\right) + \left[\mathfrak{Q}\left(a^0\right)\psi_\beta\left(\cdot, R^{n^k}\right)\right]\left(x^0\right) - \beta\psi_\beta\left(x^0, R^{n^k}\right)$$
$$= \Delta\left(x^0, a^0\right) + \varepsilon^{n^k}, \quad \text{where} \quad \left|\varepsilon^{n^k}\right| < \varepsilon. \quad (5.27)$$

We now construct a new strategy by setting

$$R_{a^0}^{n^k}(x) = \begin{cases} R^{n^k}(x), & x \neq x^0; \\ a^0, & x = x^0. \end{cases}$$

Then equality (5.27) can be rewritten as

$$D_{R_{a^0}^{n^k}}^{R^{n^k}}(x) = \Delta\left(x^0, a^0\right) + \varepsilon^{n^k},$$

and we have

$$D_{R_{a^0}^{n^k}}^{R^{n^k}}(x) \leq D_{R^{n^k+1}}^{R^{n^k}}(x), \qquad x \in X.$$

Because the sequence $\left\{\psi_\beta\left(x, R^n\right)\right\}$ is monotone and bounded,

$$0 \leq \psi_\beta\left(x, R^{n^k+1}\right) - \psi_\beta\left(x, R^{n^k}\right) \to 0, \qquad k \to \infty, \quad x \in X.$$

Recall (Definition 5.40) that the transition probability $\mathrm{P}(x, s, C; R)$ of the homogenous Markov process under control strategy R has holding time intensities that are bounded by Λ. Then

$$
\psi_\beta\left(x^0, R^{n^k+1}\right) - \psi_\beta\left(x^0, R^{n^k}\right)
$$

$$
= \int_0^\infty e^{-\beta s}\left[T^s\left(R^{n^k+1}\right)D_{R^{n^k+1}}^{R^{n^k}}(\cdot)\right](x^0)\,ds
$$

$$
\geq \int_0^\infty e^{-\beta s}\,\mathsf{P}\left(x^0, s, \{x^0\}; R^{n^k+1}\right)D_{R^{n^k+1}}^{R^{n^k}}(x^0)\,ds
$$

$$
\geq D_{R^{n^k+1}}^{R^{n^k}}(x^0)\int_0^\infty e^{-(\beta+\Lambda)s}\,ds
$$

$$
= \left[\Delta\left(x^0, a^0\right) + \varepsilon^{n^k}\right]\frac{1}{\beta+\Lambda}.
$$

Because $\varepsilon > 0$ is arbitrary, we have a contradiction. The proof is complete. $\square$

Remark 5.52. If R_β is ψ-optimal, then the optimal reward $\psi_\beta(x, R_\beta)$ satisfies for all $x \in X$ the equation

$$
\beta\psi_\beta(x, R_\beta) = \max_{a \in A(x)}\left\{r(x, a) + \left[\mathfrak{Q}(a)\psi_\beta(\cdot, R_\beta)\right](x)\right\}, \qquad (5.28)
$$

and vice versa; any strategy $R \in LS_D^d$ that maximizes for all $x \in X$ the expression in the curved brackets in Equation (5.28) is ψ-optimal.

On the other hand, if for some strategy $\pi \in LS_{PM}^d$ the value function $\psi_\beta(x, \pi)$ satisfies equation (5.28) for all $x \in X$, then for every stationary strategy $R \in LS_D^d$

$$
D_R^\pi(x) \leq 0, \qquad x \in X,
$$

and due to Theorems 5.47 and 5.45, the strategy π is then ψ-optimal. $\odot$

5.2.6 Existence of average-optimal Markov strategies in the class of local deterministic strategies

In this section we find conditions that guarantee that there exists a ϕ-optimal strategy in LS_D^d.

Theorem 5.53. Assume that there exist a constant g and a measurable bounded function $v(x)$ on X such that

$$g = \max_{a \in A(x)} \left\{ r(x, a) + \left[\mathfrak{Q}(a) v(\cdot) \right](x) \right\}. \tag{5.29}$$

Then a strategy $R^\star \in LS_D^d$ that maximizes the expression in the waved brackets in Equation (5.29) for every $x \in X$ is ϕ-optimal, and

$$\phi(R^\star; x) \equiv g. \tag{$\odot$}$$

Proof: Let $\pi = \left(R^1, t^1, R^2, t^2, \ldots \right)$ be an arbitrary admissible local pure Markov strategy in LS_D^d. Consider the following functional of the controlled process (ξ, π):

$$\Phi_x^\pi(t) = \mathsf{E}_x^\pi \left\{ \sum_{k=1}^n \int_0^{t^k - t^{k-1}} \left[T^s(R^k) r(\cdot, R^k(\cdot)) \right] \left(\eta(t^{k-1}) \right) ds + \right.$$

$$+ \int_0^{t - t^n} \left[T^s(R^{n+1}) r(\cdot, R^k(\cdot)) \right] \left(\eta(t^n) \right) ds +$$

$$\left. + \left[T^{t - t^n}(R^{n+1}) v(\cdot) \right] \left(\eta(t^n) \right) \right\}, \tag{5.30}$$

where $\eta(0) = x$, and $t \in [t^n; t^{n+1})$.

Assume that interchanging differentiation with respect to t and expectation in expression (5.30) is justified. Then, according to (5.29),

$$\frac{d}{dt}\Phi_x^\pi(t) = \mathsf{E}_x^\pi \left\{ \left[T^{t-t^n}\left(R^{n+1}\right)\left(r(\cdot, R^k(\cdot)) + \mathfrak{Q}\left(R^{n+1}\right)v(\cdot)\right)\right]\left(\eta\left(t^n\right)\right)\right\}$$

$$\leq g.$$

Integrating the last relation with respect to t from 0 to T we obtain

$$\Phi_x^\pi(T) - \Phi_x^\pi(0) \leq gT.$$

But $\Phi_x^\pi(0) = v(x)$ (which is by assumption a bounded function) and, consequently,

$$\phi(x, \pi) = \lim_{T\to\infty}\frac{1}{T}\Phi_x^\pi(T) \leq g, \qquad x \in X.$$

On the other hand,

$$\frac{d}{dt}\Phi_x^{R^\star}(t) \equiv g$$

and, consequently,

$$\phi(x, R^\star) \equiv g.$$

$\cdot$ Thus the strategy $R^\star$ is ϕ-optimal.

We finally have to prove that interchanging differentiation and expectation in expression (5.30) is justified. Denote the expression in brackets in (5.30) by $U^\pi(t)$. It is sufficient to prove that for sufficiently small $\Delta t > 0$ such that $t + \Delta t \in [t^{n-1}, t^n)$ holds

$$\frac{1}{\Delta t}\left|U^\pi(t + \Delta t) - U^\pi(t)\right| < c < \infty.$$

We have (recall $0 \le r(x, a) \le l < \infty$)

$$\frac{1}{\Delta t} \left| U^\pi(t + \Delta t) - U^\pi(t) \right|$$

$$= \frac{1}{\Delta t} \left| \int_t^{t+\Delta t} \left[T^s (R^{n+1}) r(\cdot, R^{n+1}(\cdot)) \right] \left(\eta(t^n) \right) ds + \right.$$

$$\left. + \left[T^{t+\Delta t - t^n} (R^{n+1}) v(\cdot) \right] \left(\eta(t^n) \right) - \left[T^{t-t^n} (R^{n+1}) v(\cdot) \right] \left(\eta(t^n) \right) \right|$$

$$\le \frac{1}{\Delta t} \left| l\Delta t + \left[T^{t-t^n} (R^{n+1}) \left(I + \mathfrak{Q}(R^{n+1})\Delta t + \frac{\mathfrak{Q}^2(R^{n+1})\Delta t^2}{2!} + \right. \right. \right.$$

$$\left. \left. \left. + \ldots \right) v(\cdot) \right] \left(\eta(t^n) \right) - \left[T^{t-t^n} (R^{n+1}) v(\cdot) \right] \left(\eta(t^n) \right) \right|$$

$$\le l + \frac{1}{\Delta t} \left| \left[T^{t-t^n} (R^{n+1}) \left(\mathfrak{Q}(R^{n+1})\Delta t + \frac{\mathfrak{Q}^2(R^{n+1})\Delta t^2}{2!} + \right. \right. \right.$$

$$\left. \left. \left. + \ldots \right) v(\cdot) \right] \left(\eta(t^n) \right) \right|$$

$$\le l + \left(\left\| \mathfrak{Q}(R^{n+1}) \right\| + \frac{\left\| \mathfrak{Q}(R^{n+1}) \right\|^2}{2!} + \ldots \right) \|v\|$$

$$\le l + \|v\| \left(e^{\left\| \mathfrak{Q}(R^{n+1}) \right\|} - 1 \right) < c < \infty.$$

Here I is the identity operator; recall that for given $R \in LS_D^d$ the operator $\mathfrak{Q}(R)$ is a bounded. The proof is complete. $\square$

Thus Theorem 5.53 offers sufficient conditions for a stationary strategy $R^\star$ to be ϕ-optimal and provides in principle a method to find an optimal strategy. But these sufficient conditions are given as existence conditions for bounded solutions of nonlinear equations of Bellman type. We therefore describe a further sufficient condition that guarantees the existence of a bounded solutions of Equation (5.29).

For discount parameter $\beta > 0$ let R_β be a stationary ψ-optimal strategy and x^0 be some fixed state in X. We introduce following abbreviations:

$$U_\beta(x) = \psi_\beta(x, R_\beta) - \psi_\beta(x^0, R_\beta),$$
$$g_\beta = \beta\psi_\beta(x^0, R_\beta).$$

Then Equation (5.28) can be rewritten as

$$g_\beta = \max_{a \in A(x)} \left\{ r(x, a) + \left[\mathfrak{Q}(a)\psi_\beta(x, R_\beta)\right](x) - \beta U_\beta(x) \right\},$$

which is (because of $\left[\mathfrak{Q}(a)\psi_\beta(x, R_\beta)\right](x) = 0$),

$$g_\beta = \max_{a \in A(x)} \left\{ r(x, a) + \left[\mathfrak{Q}(a)U_\beta(\cdot)\right](x) - \beta U_\beta(x) \right\}. \qquad (5.31)$$

Theorem 5.54. Assume that for some sequence $\beta^n \to 0$ holds the inequality

$$\left| U_{\beta^n}(x) \right| \leq N < \infty, \qquad x \in X.$$

Then there exist a constant g and a measurable bounded function $v(x)$ such that relation (5.29) holds. $\qquad \odot$

Proof: Due to Tikhonov's theorem on products of compact sets, the set of measurable functions on a measurable space with values in some compact space is compact in the topology of pointwise convergence. It follows that from the sequence $\{U_{\beta^n}(x)\}$ we can select a convergent subsequence $\{U_{\beta^{n'}}(x)\}$ that converges to some measurable bounded function $U(x)$ pointwise on X. Because $|g_\beta| < l$ holds, we find that the associated sequence $\left\{ g_{\beta n'} \right\}$ converges to some limit g. Using the Lebesgue's integration theorem, from (5.31) we have (5.29). □

There is a connection between ψ-optimal stationary strategies R_β for given β and ϕ-optimal strategy $R^\star \in LS_D^d$, which is formulated in the following theorem.

Theorem 5.55. Let the assumption of Theorem 5.54 hold, especially $\beta^n \to 0$. Then the limit of every convergent sequence $\left\{ R_{\beta n'} \right\} \subset \left\{ R_{\beta n} \right\}$ is a ϕ-optimal strategy. The existence of a convergent sequence is guaranteed by Tikhonov's theorem. ⊙

We finally formulate in probabilistic terms sufficient conditions that imply that the basic assumptions of Theorem 5.54 are fulfilled.

Corollary 5.56. The assumptions of Theorem 5.54 are fulfilled if the following holds.

1) For every strategy $R \in LS_D^d$ the transition probability $\mathsf{P}(t; x, C; R)$ of the homogeneous Markov process controlled by strategy R has a density $\mathsf{p}(t; x, y; R)$ with respect to some finite measure μ on $(X, \mathfrak{X})$, i.e.,

$$\mathsf{P}(t; x, C; R) = \int_C \mathsf{p}(t; x, y; R)\mu(dy).$$

2) The Markov process controlled by strategy R has an ergodic distribution $\mathsf{P}(C; R)$ with density $\mathsf{p}(y; R)$ with respect to μ, i.e.,

$$\lim_{t\to\infty} \mathsf{P}(t; x, C; R) = \mathsf{P}(C; R) = \int_C \mathsf{p}(y; R)\mu(dy).$$

3) Doeblin's condition (see [Doo53, p. 256]) holds, i.e., the convergence $\mathsf{p}(x, t, y; R) \to \mathsf{p}(y; R)$ for $t \to 0$ is exponential. ⊙

Remark 5.57. If the existence of ψ-optimal stationary strategy was proved for finite action space A and bounded reward function $r(x, a)$, then in general, to prove the existence of a ϕ-optimal stationary strategy, we have to assume some strong ergodicity of the controlled Markov process [BSL90].$\odot$

Chapter 6

CONNECTIONS WITH OPTIMIZATION OF RANDOM FIELD IN DIFFERENT AREAS

In this chapter we describe selected examples from different fields of applications where classical Markov fields or Gibbs fields are successfully applied to modeling and optimization of large systems with graph structured state space.

These models are related to the models that we summarized under the heading of *time dependent Markov or semi-Markov random fields* in the previous chapters. We believe that these examples will enlighten the principles that guided us through the previous chapters although — as will be seen soon — there are, in any case, differences to our modeling principles. We discuss these differences in any case and so further clarify the ways to deal with the large systems we have in mind.

In Section 6.1 we reconsider the model for diffusion of knowledge and technologies that we already presented in previous chapters; see Example 3.41. In contrast to our approach in the literature, there is usually optimization on the basis of the overall state description prevalent, but usually no time parameter is considered.

In Section 6.2 we discuss by very limited examples the connection to image recognition, image segmentation, and restoration. Here we do not go into any technical details, but use the proposed problems to clarify the necessity for parameter estimation for random fields that serve as models for image generation.

An estimation procedure that uses an iterative stochastic approximation is described in Section 6.3. Although there is no generic time development in the problem setting, nevertheless in this iteration procedure, a stochastic process with one-dimensional marginals that are random fields emerges.

In Section 6.4 we describe an approach developed by Belyaev [Bel00] for the classification and recognition of colored pictures. The classification mechanism is not specified in detail there, but he considered the probabilities for successful classification and for misclassification of the final decision to be known. Assuming locality properties similar to those we used to investigate throughout this text, he arrives at limit theorems for the classification procedures. Again, there is no generic time development in the problem setting but due to the steps

- observation and estimation,
- classification and assessment of the procedure,

there occurs an iteration where a localized kernel plays the central role. So again, an iteration with one-dimensional marginals that are random fields emerges.

In Section 6.5 the microscopic behavior of financial markets is discussed along the lines of [Voi03, Chapter 8]. Discrete populations of agents buy and sell units of one stock and by observing the prices over time, they interact by incorporating the observed behavior of other agents into their decision process. If this decisions depend on observing the agents in their neighborhood only, we can describe the sequential trading decisions by local and synchronous processes. But in general this is not the case because the price process usually depends on the global behavior of all agents and is incorporated into the decisions usually. We discuss the relations of different models to our modeling principles.

6.1 DIFFUSION OF KNOWLEDGE AND TECHNOLOGIES

In Example 3.41 we discussed the problem of competing technologies used by different companies and pointed out the relation to models of classical statistical physics. The relevant model for the case of only two different technology levels is the stochastic Ising model. We describe some details here and discuss further model specifications.

Definition 6.1. Consider an undirected graph $\Gamma = (V, B)$ and local state spaces $X_i = \{+1, -1\}$ for all nodes $i \in V$. The global states $x = x_V \in X = \{+1, -1\}^V$ are called *configurations*, and considered as functions on Γ with values ± 1. A function

$$U_V \colon X \to \mathbb{R}, \quad U_V(x_V) = h \sum_{k \in V} x_k + \beta \sum_{k \in V} \sum_{k' \in N(k)} x_k x_{k'}, \quad (6.1)$$

determines the *energy of configuration* x_V. A physical system with phase space $X = \{+1, -1\}^V$, which consists of all configurations x_V and an energy function $U_V(x_V)$ for the configurations defined by (6.1), is called an *Ising model* (on a finite graph). ⊙

Under stationary conditions the probability of configuration x_V is

$$\mathsf{P}_V(x_V) = Z_V^{-1} \exp\{-U_V(x_V)\}, \quad (6.2)$$

where $Z_V = \sum_{x_V \in X} \exp\{-U_V(x_V)\}$ is the *partition function* (normalizing constant). The distribution (6.2) is a Gibbs distribution for the corresponding Ising model.

Consider a random field $\xi \colon (\Omega, \mathcal{F}, \mathrm{Pr}) \to (X, \mathfrak{X})$ with distribution given in (6.2). (Here $\mathfrak{X} = 2^X$.)

For all $K \subseteq V$ the expectations (under the distribution from (6.2)) $\mathsf{E} \prod_{k \in K} \xi_k$ are called the correlation functions or moments of the distribution (6.2).

For every $K = \{k_1, \ldots, k_n\} \subset V$ the joint distribution

$$\mathsf{P}_K\{x_{k_1}, \ldots, x_{k_n}\} =: \Pr\{\xi_{k_1} = x_k, \ldots, \xi_{k_n} = x_{k_n}\},$$

where $\{x_{k_1}, \ldots, x_{k_n}\} \in \{+1, -1\}^K$ is an arbitrary configuration on K, can be computed via the correlation functions. It can be shown [MM85]:

$$\mathsf{P}_K\{x_i \colon i \in K\} = \frac{(-1)^k}{2^n} \sum_{H \subseteq K} \left(\prod_{j \in K-H} x_j \right) \mathsf{E}\left[\prod_{i \in H} \xi_i \right],$$

where k is the number of values $x_i = -1, i \in K$.

If $\mathsf{E} \prod_{k=1}^{n} \xi_k$ is not known, then we need a statistical estimate of the respective moments such that we can then compute an estimate for the joint distribution of $(\xi_k, \ k \in K)$.

In statistical physics, the graphs under consideration are finite regular sublattices of $\mathbb{Z}^n$ for some n. The thermodynamic limit is then obtained by letting the sublattice grow to $\mathbb{Z}^n$. This limit is used as approximation for large graph structured systems. The problem that arises is that of existence and uniqueness of the limiting distribution. As a simple example, consider in the conditional probabilities in Equation (3.6) with the following specification

$$q_u\big(y_k \mid x_{N(k)}\big) = \frac{\exp\left[-\beta \sum_{j \in N(k)} y_k x_j\right]}{\sum_{y \in X} \exp\left[-\beta \sum_{j \in N(k)} y x_j\right]},$$

where y_k and x_j take values ± 1, and $\beta > 0$.

Under this condition there exists some β^* such that over the infinite lattice $\mathbb{Z}^n$, a unique invariant and ergodic distribution exists if $0 < \beta \leq \beta^*$ [Lig85]. But even in this case, to determine equilibria in explicit form is not an easy task. For a special result see O. N. Stavskaya [Sta71] where the conditional probabilities $q_k\big(y_k \mid x_{N(k)}\big)$ are equal for all k and the states at different vertices in any moment change independently of each other.

Example 6.2. We reconsider Example 3.41 concerning the optimal policy of a company when there are several firms with competing technologies. Assume that the i-th company chooses between only two different levels of technology, denoted ± 1, and assume that there is no control. If $r_i(j)$ is the reward from selling during one time period products that are produced on level j by the i-firm and if $f_i(j)$ is the steady state probability that the i-th firm will produce on level j with $i = 1, \ldots, n$, and $j = \pm 1$, then the mean reward of firm i in one period is

$$\mathsf{E}\, r_i = f_i(1)r_i(1) + f_i(2)r_i(2). \tag{6.3}$$

The form of the transition probabilities (3.6) in this example with finite interaction graph implies positive recurrence of the Markov chain and therefore (6.3) is the long-run time average as well. The $f_i(j)$ depend on the neighborhood structure of the graph and additionally on (the temperature) β. For a fixed reward function we therefore can optimize the stationary reward per time unit in β over some prescribed interval. This would be subject of classical global optimization.

Local optimization in the sense developed in the previous chapters would require a different approach. We have to open the opportunity to the firms to control some additional parameter that has to be introduced into the system and would give rise to (local) costs for, e.g., using new machines, applying more expensive control and maintenance, and so on.

A way to introduce local controls was sketched in Example 3.41, referring to local temperatures at different nodes.   ⊙

In this and similar ways, we can apply the results of Section 3 to Ising-like systems.

An interesting related model of evolutionary driven industrial dynamics where firms use different technologies and may get knowledge of up to then hidden technological routines of other firms is described in [KK92]. The model is used as a basis of simulating the market development under these sort of information diffusion. The authors do not rely on neighborhood structures but seemingly, this is a promising topic to be included explicitly into the picture.

Another simulation study of innovation in complex technology spaces that is closer to our neighborhood concept is performed by [SV02]. Their mathematical model relies on the concept of percolation processes, which are graph structured models developing in discrete time in their setting. For further related research, see the list of references there.

6.2   IMAGE RECOGNITION

There are many approaches to the problem of images recognition and identification for digital images. One approach is based on the choice of (different) probability models for the representation of the images. Assuming we have some a priori information about the true image, this should have strong influence on the model selection process. And the choice of the model mainly determines the solution method for the problem. The approach we have in mind is to use random field models as representations of the images. There is much literature dealing with these methods and we will not directl contribute herey. Our intention is to give the reader an introduction as to what the directions are, and where the random fields apply. For more details, see especially the book of Winkler [Win95] and the many references given there. We determine then adequate parameters of the random field's distribution such that varying the parameters yields a versatile class of random fields and their distributions to fit an observed image.

A standard assumption that is justified in many areas of image recognition or segmentation is that the random fields underlying the observation are Gibbs fields over some graph $\Gamma = (V, B)$, where V is usually some finite subset of $\mathbb{Z}^2$ with B often the natural neighborhood graph. Then natural parameters are the potential or the energy of the distribution; see Definition 3.8. While the potential may be a complicated high dimensional parameter, the energy is a function of the configuration only, and so is easier to access. Often the energy is defined in a way that it still depends on an additional single parameter, usually called temperature, such that with the notation of Definition 3.8 for a given (Gibbs) potential u and any subgraph $J \in V$, the *energy* $U(x_J)$ of the (local) configuration $x_J \in \underset{i \in J}{\times} X_i$ under given temperature $T \in \mathbb{R}$ is

$$U(x_J) = \sum_{\substack{C \subseteq J \\ C \in \mathcal{C}}} \frac{1}{T} u_C(x_C).$$

If the potential u is not specified, we are faced with nonparametric estimation, while assuming that the potential u is prescribed we can perform parametric estimation, e.g., apply Maximum Likelihood methods.

Problems of image recognition and classification in the setting described above then are as follows.

Assume an *ideal* or *standard* image exists, the distribution of which is known to us, and we have made observations (pictures). We have to decide whether the observations are realization of the standard image. This can be done by comparing the estimated potential or energy with that of the standard image. So we have to estimate either the potential or the energy or often, the temperature only. It turns out that in many cases, the problems can be converted into a question on maximizing or minimizing the energy or the temperature. There are many methods to perform this by using stochastic optimization algorithms.

In practice there are many variants of this problem:

• There are several *standard* images and we have to decide to which our observation fits best.

• We do not have prescribed *standard* images but have to extract by statistical methods form a first set of data the parameters of interest. And then proceed as above. This occurs often in connection with machine learning.

• The standard image may be perturbated by noise (with either known or unknown distribution).

6.3 PARAMETER ESTIMATION

In this section we sketch a simple example of how to estimate the parameters of random fields in connection with the image recognition problems described in the previous subsection. In our description we follow [Gim90, GZ89, Zal91]. For a short introduction into the basic principles of maximum likelihood estimation in the context of random fields, see [Win95, Sections 13]; for more information on spacial maximum likelihood estimation, see Section 14 there.

We consider a random field

$$\xi\colon (\Omega, \mathcal{F}, \mathrm{Pr}) \to (X, \mathfrak{X}) = \left(\underset{i \in V}{\times} X_i, \bigotimes_{i \in V} \mathfrak{X}_i \right)$$

over $\Gamma = (V, B)$ which is a Gibbs field with distribution

$$\begin{aligned}
\mathrm{Pr}(\xi = x) &= \mathsf{P}_T(x) \\
&= Z^{-1} \exp\left(-\frac{1}{T} U(x) \right) \qquad (6.4) \\
&= Z^{-1} \exp\left(-\sum_{\substack{C \subseteq V \\ C \in \mathcal{C}}} \frac{1}{T} u_C(x_C) \right),
\end{aligned}$$

where the set $u = (u_C \colon C \subseteq V)$ of functions $u_C \colon \underset{i \in C}{\times} X_i \to \mathbb{R}$ is a *Gibbs potential*, i.e., u_C vanishes for all $C \subseteq V$ that are not a clique. Z is the partition function (normalization constant).

The potential $\frac{1}{T} u_C(\cdot)$ in this example is therefore parameterized by the (*generalized*) *temperature* T, or by $\lambda = \frac{1}{T}$, and so are the energy $U(\cdot)$ and the distribution P_T. Note that due to the finite underlying graph for our random field, the notion of a critical temperature does not occur, so from the strict positivity it follows that the conditional distributions (specifications, see Definition 3.6) uniquely determine the Gibbs field. Our problem is in principle a standard one:

For an observation $x \in X$ find

$$T_{\max}(x) = \arg\max_T P_T(x), \qquad (6.5)$$

i.e., find the maximum likelihood estimator for the temperature parameter T for an observed configuration of the field ξ.

We need the following result.

Theorem 6.3 (see [Gim90, GZ89, Zal91]). Consider the class of Gibbs fields with prescribed Gibbs potential from (6.4). Then for every configuration $x \in X$ the function

$$T \longrightarrow P_T(x)$$

has a unique maximizer $T_{\max}(x) = \arg\max_T P_T(x)$ and it holds:

$$U(x) > \frac{\sum_{y \in X} U(y)}{|X|} \implies (T_{\max}(x) = \infty) \qquad (6.6)$$

$$U(x) \leq \frac{\sum_{y \in X} U(y)}{|X|} \implies \left(U(x) = \mathsf{E}_{T_{\max}(x)} U(\cdot) \right), \qquad (6.7)$$

where U is the energy function, $E_{T_{\max}(x)}$ is expectation under $\mathsf{P}_{T_{\max}(x)}$, i.e., integral under the Gibbs distribution with potential $u(\cdot)$ and with temperature $T_{\max}(x)$. $\qquad \odot$

From Theorem 6.3 there exists a unique solution to (6.5), which we compute by stochastic programming. It follows further from this theorem that for estimating the unknown parameter T of an image (i.e., a Gibbs field), we must compute the expectation of the energy function $U(\cdot)$ under the true distribution with parameter $T_{\max}(x)$. This is often a hard computational problem because of the size of the configuration space. Therefore approximation of this expectation is needed, and then using this in the solution of the optimization problem. The problem arises to prove convergence of the procedure.

For simplicity we set $\lambda = \frac{1}{T}$ and solve (6.5) for λ:

$$\lambda_{\max}(x) = \arg\max_{T} P_\lambda(x).$$

The convergence of our solution procedure follows from Theorem 6.4 below, which deals with the following approximation scheme.

Let $(\Omega, \mathcal{F}, \Pr)$ be a probability space and $\Xi = [a, b]$ with $0 \leq a < b \leq \infty$. Let the function

$$F \colon \Xi \times \Omega \to \mathbb{R}$$

be $\mathcal{F}$-measurable for each fixed $x \in \Xi$. We consider the maximization problem for the function $\Phi(x) = \mathsf{E}\, F(x, \omega)$, $x \in \Xi$. We define the random sequence

$$x^{k+1} = \text{proj}_\Xi\left(x^{k+1} + \rho_k \xi^k\right), \qquad k = 0, 1, \ldots, \tag{6.8}$$

where

- $\xi^k = \xi^k(\omega)$, $k \in \mathbb{N}$, is a sequence of random variables (stochastic gradients), which satisfies $\mathsf{E}\left[\xi^k \mid x^0, x^1, \ldots, x^k\right] = \frac{d}{dx}\Phi(x^k)$ for all $x^k \in \Xi$, $k \in \mathbb{N}$,
- $\text{proj}_\Xi(\cdot)$ is the projection operator , which is the identity on Ξ and maps all $y < a$ on a, and all $y > b$ on b,
- ρ_k, $k \in \mathbb{N}$, is a sequence of non-negative numbers,
- x^0 is some initial value in Ξ.

Theorem 6.4 (see [Erm76]). Assume that the function $\Phi(x)$ is concave and differentiable on Ξ and the following conditions are satisfied:

1) $\mathsf{E}\left[\,|\xi^k|\mid x^0, x^1, \ldots, x^k\right] < c$, for some $c < \infty$;
2) $\sum_{k=0}^{\infty} \rho_k = \infty$, $\sum_{k=0}^{\infty} \rho_k^2 < \infty$.

Then

$$\Pr\left\{\lim_{k\to\infty}\left|x^k - x^\star\right| = 0\right\} = 1,$$

where $x^\star \in \Xi$ is a maximizers of $\Phi(x)$. If $\Phi(x)$ is strictly concave, then $x^\star$ is uniquely determined. ⊙

We apply Theorem 6.4 to compute the required maximum likelihood estimator. First we compute for fixed configuration (observation) $x \in X$ the derivative with respect to λ of $P_\lambda(x)$:

$$\frac{\partial P_\lambda(x)}{\partial \lambda} = \frac{-U(x)e^{-\lambda U(x)}\sum_{y\in X}e^{-\lambda U(y)} + e^{-\lambda U(x)}\sum_{y\in X}U(y)e^{-\lambda U(y)}}{\left(\sum_{y\in X}e^{-\lambda U(y)}\right)^2}$$

$$= P_\lambda(x)\left[\mathsf{E}_\lambda U(\cdot) - U(x)\right], \qquad (6.9)$$

where E_λ is expectation with respect to P_λ.

If for a fixed observation $x \in X$ condition (6.6) of Theorem 6.3 is satisfied, then it is easy to see that the maximum of $P_\lambda(x)$ is attained at $\lambda = 0$. If on the other hand, condition (6.7) of Theorem 6.3 is satisfied, then the point λ for which $\mathsf{E}_\lambda U(\cdot) = U(x_s)$ holds is stationary for $P_{(\cdot)}(x)$ according to (6.9). This is immediate from strict positivity of the random field's counting density. So the problem to find $\max_\lambda P_\lambda(x)$ is equivalent to computation of the root of the function $R(\lambda) = \mathsf{E}_\lambda U(\cdot) - U(x)$.

We apply the iterative procedure (6.8) to compute the optimal $\lambda = \lambda_{\max}(x)$ for observation $x \in X$, which is explicitly

$$\lambda^{k+1} = \lambda^k + \rho_k \xi^k.$$

The random variables ξ^k are computed as follows:

$$\xi^k = \langle U(x)\rangle_k - U(x),$$

where $\langle U(x)\rangle_k$ is an estimator for $\mathsf{E}_{\lambda^k} U(\cdot)$:

$$\langle U(x)\rangle_k = \frac{1}{N}\sum_{i=1}^{N} U(v^i),$$

where $v^i, i = 1, \ldots, N$, is a sequence of random fields with distribution P_{λ^k}. These random fields are simulated in course of the iterations at step k and $\langle U(x) \rangle_k$ is therefore an estimator for $\mathsf{E}_{\lambda^k} U(\cdot)$. It is proved in [Zal91] that $\langle U(x) \rangle_k$ is indeed a consistent estimator for $\mathsf{E}_{\lambda^k} U(\cdot)$.

From Theorem 6.4, it follows that

$$\Pr \left\{ \lim_{k \to \infty} \left| \lambda^k - \lambda_{\max}(x) \right| = 0 \right\} = 1.$$

Using classical results of stochastic approximation, it can be proved that under some additional technical assumptions for the sequence

$$\sqrt{k} \left(\lambda^k - \lambda_{\max}(x) \right), \qquad k \to \infty,$$

a central limit theorem holds [NH73, p. 151], [KKN80].

6.4 CLASSIFICATION AND RECOGNITION PROBLEMS

In this section we describe a method developed by Belyaev [Bel00] to identify and classify colored pictures. We assume a fixed colored picture is given but the observation of the picture is perturbed by noise. The picture is defined pixel by pixel on a finite regular graph $V = \{ (i,j) \colon 1 \leq i \leq n, 1 \leq j \leq m \} \subseteq \mathbb{Z}^2$. The edges of the graph are given by the natural nearest neighborhoods $N(i,j) = \{ (i, j+1), (i+1, j), (i, j-1), (i-1, j) \}$ for interior points of V and the restricted similar neighborhoods for boundary points. So the picture is a function

$$c \colon V \longrightarrow K,$$

where K is the set of admissible colors, $|K| = k_0$. But instead of the deterministic field, we observe a random field of colors

$$\xi \colon (\Omega, \mathcal{F}, \Pr) \to (X, \mathfrak{X}) = \left(K^V, \mathfrak{K}^V \right)$$

where $\mathfrak{K} = 2^K$.

Belyaev does not make assumptions on the structure of the noise, so we have applicability of his results to general observations. He assumes that a certain classification procedure is applied to recover the original picture, i.e., based on the noisy observation $x = \xi(\omega) \in X$ to associate to each pixel either a color from the set K, or the symbol $\phi \notin K$, which indicates that assignment of a color is not possible. We denote by $K_0 = K \cup \{\phi\}$ and refer to ϕ as a (generalized) color as well. The classification procedure can be a randomized algorithm. The main assumption for us to assess the quality of the classification is on the probability to correctly assign the colors, or on the error probabilities. Belyaev prescribes that the interplay of noise and (randomized) classification procedure results in a structure of the classification probabilities that strongly resembles our properties of synchronization and localization that occurred throughout the previous chapters; see, e.g., Definition 3.10.

Assumption 6.5 (Conditional independence and local structure of classification). The result of the classification based on the observation is a random field

$$C^\bullet : (\Omega, \mathcal{F}, \mathrm{Pr}) \to (X_0, \mathfrak{X}_0) = \left(K_0^V, \mathfrak{K}_0^V\right).$$

such that for any pair of pixels $\{(i_h, j_h) : h = 1, 2\}$ holds:
Given $x \in X$ and colors $\{l_h \in K_0, k_h \in K : h = 1, 2, \}$

$$\mathrm{Pr}\{C^\bullet(i_1, j_1) = l_1, C^\bullet(i_2, j_2) = l_2 \mid c(i_h, j_h) = k_h, \; \xi = x, \; h = 1, 2\}$$

$$= \prod_{h=1,2} \mathrm{Pr}\left\{C^\bullet(i_h, j_h) = l_h \mid c(i_h, j_h) = k_h, \; \xi_{N(i_h,j_h)} = x_{N(i_h,j_h)}\right\},$$

i.e., these transition probabilities are local and pairwise synchronized. $\odot$

Assumption 6.6. The probabilities for classification

$$\mathrm{Pr}\left\{C^\bullet(i, j) = l \mid c(i, j) = k, \; \xi_{N(i_h,j_h)} = x_{N(i,j)}\right\} =: p_{kl}\left(x_{N(i,j)}\right),$$

$k \in K$, $l \in K_0$ are known functions of the observations $x_{N(i,j)}$.   ⊙

The matrix

$$\mathbb{P}\Big(x_{N(i,j)}\Big) = \Big(p_{kl}\big(x_{N(i,j)}\big) : k \in K, l \in K_0\Big),$$

which consists of the probabilities of all possible classifications, has k_0 rows and (k_0+1) columns. We fix the last column to refer to ϕ. Consider the transposed $(k_0 + 1 \times k_0)$ matrix $\mathbb{P}^T\Big(x_{N(i,j)}\Big)$.

For each $h \in K_0$, delete the h-th row of the transposed matrix to obtain the $(k_0 \times k_0)$ matrix $\mathbb{P}_h\Big(x_{N(i,j)}\Big)$. So $\mathbb{P}_\phi(\cdot)$ is obtained by deleting the last row for each local observation $x_{N(i,j)}$.

Assumption 6.7. Under any observation $x_{N(i,j)}$ the classification probability matrices $\mathbb{P}_h\Big(x_{N(i,j)}\Big)$, $h \in K_0$, are of rank k_0.

We define the respective inverse matrices by

$$\mathbb{P}_h\Big(x_{N(i,j)}\Big)^{-1} = \Big(p_h^{lk}\big(x_{N(i,j)}\big) : l \in K_0 - \{h\}, k \in K\Big)$$

for each $h \in K_0$.   ⊙

Let us denote the realization of the randomized classification procedure

$$C^\bullet : (\Omega, \mathcal{F}, \mathrm{Pr}) \to (X_0, \mathfrak{X}_0) = \big(K_0^V, \mathfrak{K}_0^V\big)$$

by

$$c^\bullet : V \to (X_0, \mathfrak{X}_0) = \big(K_0^V, \mathfrak{K}_0^V\big),$$

which yields for each pixel $(i, j) \in V$ some classification $c^\bullet(i, j) \in K_0^V$.

Assessing the goodness of fit for a classification would need in principle comparison with the underlying fixed picture (the colors of the pixels

$c(i,j))$ with the approximated classifications $c^\bullet(i,j)$, obtained from the observation $x(i,j)$.

Recall the use of indicator functions; see page 10. Let

$$n^\bullet_{kl} := \sum_{(i,j)} \mathbb{1}\big(c(i,j) = k\big)\mathbb{1}\big(c^\bullet(i,j) = l\big). \tag{6.10}$$

So for $k \neq l$ we see that with increasing $n^\bullet_{kl}$ the goodness of the classification of pixel (i,j) decreases. But the values $n^\bullet_{kl}$ are unobservable. Therefore we have to find some unbiased estimator for the $n^\bullet_{kl}$ and to study the asymptotic properties of the estimators.

Starting from the observation that under Assumptions 6.5 and 6.6 holds

$$\mathsf{E}\Big[\mathbb{1}\big(C^\bullet(i,j) = l\big)\Big] = \sum_{k \in K} \mathbb{1}\big(c(i,j) = k\big)p_{kl}\big(x_{N(i,j)}\big),$$

Belyaev [Bel00] derived the following unbiased estimators for the true color of pixel (i,j):

$$\hat{\mathbb{1}}_h\big(c(i,j) = k\big)$$
$$= \sum_{l \in K_0 - \{h\}} p_h^{lk}\big(x_{N(i,j)}\big)\mathbb{1}\big(C^\bullet(i,j) = l\big), \qquad k \in K. \tag{6.11}$$

In general, we may use weighted unbiased estimators of the following form [Bel00]:

$$\hat{\mathbb{1}}_{w_k}\big(c(i,j) = k\big) = \sum_{h \in K_0} w_{kh}\hat{\mathbb{1}}_h\big(c(i,j) = k\big), \qquad k \in K,$$

where $w_k = \{w_{k,h} : h \in K_0\}$, and for all $k \in K$ $w_{k,h} \geq 0$, $h \in K_0$, and $\sum_{h \in K_0} w_{kh} = 1$.

Now let $N^\bullet_{kl}$ be the random number of pixels for which $c(i,j) = k$ holds in the (unobservable) underlying picture and $C^\bullet(i,j) = l$ in the approximate random classification. Similarly to (6.10) we have

$$N^\bullet_{kl} := \sum_{(i,j)} \mathbb{1}\big(c(i,j) = k\big)\mathbb{1}\big(C^\bullet(i,j) = l\big).$$

For this number we obtain similarly the unbiased estimator $N_{kl}^{\bullet}$ is

$$\hat{N}_{kl}^{\bullet} = \sum_{(i,j)} \hat{\mathbb{1}}_{w_k}\big(c(i,j) = k\big)p_{kl}\Big(x_{N(i,j)}\Big).$$

The following asymptotic properties are proved in [Bel00].

Let $\mathcal{N} = \sum_{k\in K, l\in K_0} N_{kl}$ be the total number of pixels and define the rescaled deviations

$$D_{kl}(\mathcal{N}) = \frac{1}{\sqrt{\mathcal{N}}}\big(\hat{N}_{kl}^{\bullet} - N_{kl}^{\bullet}\big)$$

$$= \sum_{(i,j)} U_{kl}\Big(i,j,x_{N(i,j)}\Big),$$

where

$$U_{kl}\big(i,j,x(i,j)\big)$$

$$= \frac{1}{\sqrt{\mathcal{N}}}\bigg(\hat{\mathbb{1}}_{w_k}\big(c(i,j) = k\big)p_{kl}\Big(x_{N(i,j)}\Big) -$$

$$- \mathbb{1}\big(c(i,j) = k\big)\mathbb{1}\big(C^{\bullet}(i,j) = l\big)\bigg).$$

For fixed observation $x_{N(i,j)}$ from Assumptions 6.5, 6.6, and 6.7 the random variables $\Big\{U_{kl}\Big(i,j,x_{N(i,j)}\Big)\Big\}$ are independent and have mean zero. The brlow theorem then follows.

Theorem 6.8 (see [Bel00, Theorem 1]). Assume that Assumptions 6.5–6.7 hold and consider the unbiased estimators of the indicators $\mathbb{1}\big(c(i,j) = k\big)$ from (6.11).

Then for the variances $\sigma_{kl}^2(\mathcal{N})$ of the deviations $D_{kl}(\mathcal{N})$ and for the covariances $c_{k_1,l_1;k_2,l_2}(\mathcal{N})$ of the pairs $\big(D_{k_1,l_1}(\mathcal{N}), D_{k_2,l_2}(\mathcal{N})\big)$ of deviations, we have unbiased estimators of the following form.

$$\hat{\sigma}^2_{kl}(\mathcal{N})$$

$$= \frac{1}{N} \sum_{(i,j)} \left(p_{kl}\big(\mathbf{x}_{N(i,j)}\big) - 2 p^{lk}_\phi\big(\mathbf{x}_{N(i,j)}\big) p_{kl}\big(\mathbf{x}(N(i,j))\big) + 1 \right)$$

$$\times\, p_{kl}\big(\mathbf{x}_{N(i,j)}\big) \hat{\mathbb{1}}_\phi\big(c(i,j) = k\big)$$

and

$$\hat{c}_{k_1,l_1;k_2,l_2}(\mathcal{N})$$

$$= \frac{1}{N} \sum_{(i,j)} \left(p_{k_1 l_1}\big(\mathbf{x}_{N(i,j)}\big) p_{k_2 l_2}\big(\mathbf{x}_{N(i,j)}\big) \times \right.$$

$$\times \sum_{k \in K} \left(\sum_{l \in K} p_{kl}\big(\mathbf{x}_{N(i,j)}\big) p^{lk_1}_\phi\big(\mathbf{x}_{N(i,j)}\big) \times \right.$$

$$\left. \times\, p^{lk_2}_\phi\big(\mathbf{x}_{N(i,j)}\big) \right) \hat{\mathbb{1}}_\phi\big(c(i,j) = k\big) -$$

$$-\, p_{k_2 l_2}\big(\mathbf{x}_{N(i,j)}\big) p^{l_1 k_2}_\phi\big(\mathbf{x}_{N(i,j)}\big) \times$$

$$\times\, p_{k_1 l_1}\big(\mathbf{x}_{N(i,j)}\big) \hat{\mathbb{1}}_\phi\big(c(i,j) = k_1\big) -$$

$$-\, p_{k_1 l_1}\big(\mathbf{x}_{N(i,j)}\big) p^{l_2 k_1}_\phi\big(\mathbf{x}_{N(i,j)}\big) \times$$

$$\times\, p_{k_2 l_2}\big(\mathbf{x}_{N(i,j)}\big) \hat{\mathbb{1}}_\phi\big(c(i,j) = k_2\big) +$$

$$\left. +\, p_{k_1 l_1}\big(\mathbf{x}_{N(i,j)}\big) \hat{\mathbb{1}}_\phi\big(c(i,j) = l_1\big) \mathbb{1}(l_1 = l_2) \mathbb{1}(k_1 = k_2) \right)$$

for $k, k_1, k_2 \in K$, $l, l_1, l_2 \in K_0$.   ⊙

Belyaev proved [Bel00] that for $N \to \infty$, the joint distributions of the matrix of deviations $\{D_{kl}(\mathcal{N})\}_{k \in K, l \in K_0}$ converges weakly to a family of $k_0 \times (k_0+1)$-th dimensional normal distributions. Here weak convergence is generalization of weak convergence; see [Bel97, BSdL97, BSdL00].

In the spirit of the previous chapters, we can now introduce cost functions depending on incorrect classification and then introduce controls for the functions $C^\bullet(i,j)$, respectively $c^\bullet(i,j)$. Thus we may define iterative procedures to enhance the classification using methods similar to those developed in Chapter 3.

6.5 STATISTICAL MECHANICS OF FINANCIAL MARKETS

Mathematical models for the microscopic behavior of financial markets are collected from the literature and described in [Voi03, Chapter 8]. Discrete populations of agents buy and sell units of one stock and by observing the prices over time, they adjust their trading decisions and behavior. Depending on the assumptions, these agents behave Markovian in time, or act with a longer memory. With Markovian assumptions in some examples occurs a behavior that resembles the equilibrium behavior of Ising models or spin glass models, which are prototypes of Gibbsian distributed systems.

We describe in the following some details of the market models that are related to our optimization procedures in the models of the previous chapters.

There are N agents located at the vertices of some graph $\Gamma = (V, B)$ with $|V| = N$. The wealth of the agent at node i (we shall call her i from now on) is at time $t \in \mathbb{N}$

$$W_i(t) = B_i(t) + \Phi_i(t) \cdot S(t), \tag{6.12}$$

where $B_i(t)$ is the cash, $\Phi_i(t)$ the number of shares (of one stock) owned by her. $S(t)$ is the spot prize of the stock. Trading is synchronous according to the following rules.

- Agent i decides to change her $\Phi_i(t)$ by an amount (positive or negative, buy or sell) of

$$\Delta\Phi_i(t) = X_i(t) \cdot \Phi_i(t) + \frac{\gamma_i B_i(t) - \Phi_i(t) \cdot S(t)}{2\tau_i}, \qquad (6.13)$$

where γ_i, τ_i are randomly chosen but then fixed for agent i, $X_i(t) = f_i(S(t), S(t-1), S(t-2), \dots, S(t-T))$ is determined individually by a utility function f_i from the price history, usually as moving averages (with window size T) over different functionals of the prices with some additional weighting.

- In a next step, all demands and all offers are pooled to a total demand $D(t)$ and a total offer $O(t)$, and then the individual offers rescaled according to a simple mechanism which yields some value

$$\bar{\Delta}\Phi_i(t) \qquad (6.14)$$

that is transformed to agent i as the market's reaction according to her market order/offer $\Phi_i(t)$.

- This finally results in her new wealth at time $t+1$ according to

$$\Phi_i(t+1) = \Phi_i(t) + \bar{\Delta}\Phi_i(t),$$
$$B_i(t+1) = B_i(t)(1 + \varepsilon_i(t)) - \Phi_i(t) \cdot S(t)(1 + \pi).$$

$\varepsilon_i(t)$ is a very small independent perturbation, π represent transaction costs.

- Then the new price is fixed to

$$S(t+1) = S(t)\frac{\frac{1}{T+1}\sum_{i=0}^{T} D(t-i)}{\frac{1}{T+1}\sum_{i=0}^{T} O(t-i)}. \qquad (6.15)$$

In this model the interaction of the agents is a globally determined interaction, because in computing (6.14), all the offers and demands are used. Further in (6.13), the price $S(t)$ is incorporated and the latter is determined by the behavior of all agents. We conclude that

$W = \left(W(t) = \big(W_i(t) \colon i \in V \big) \colon t \geq 0 \right)$ is a spatiotemporal process that has synchronized transitions but the transitions are by no means local.

This model is transformed into a simulation model, and the outcomes are discussed and compared with results from simulating other models [Voi03, p. 219-225].

Closer to our models are the next market models that Voit describes. While in the previous model the interaction of the agents is only via an individual reaction on the series of spot prices, which means no individual neighborhoods exist, such individual neighborhoods defined by the edge set B now come into the play.

We consider the wealth process from (6.12). Trading is synchronous according to the following extended rules.

• It is additionally assumed that at one time instant, at most one stock can be bought or sold by an agent,

$$\Delta \Phi_i(t) \in \{+1, 0, -1\}.$$

• The decision is according to some individual thresholds $\xi_i^{\pm}(t)$

$$\Delta \Phi_i(t) = \begin{cases} +1, & \text{if } Y_i(t) \geq \xi_i^+(t); \\ 0, & \text{if } Y_i(t) \leq \xi_i^+(t); \\ -1, & \text{if } \xi_i^-(t) < Y_i(t) < \xi_i^-(t), \end{cases}$$

and $Y_i(t)$ can be thought to be a trading signal.

• For agent i $Y_i(t)$ is computed as follows:

$$Y_i(t) = \sum_{j \in N(i)} u(i,j) \cdot \Delta \Phi_j(t) + a\nu_i(t) + b\varepsilon(t),$$

where $a, b \in \mathbb{R}$, $u(\cdot, \cdot)$ is a Gibbsian interaction potential (see the Definition 3.8), $\nu_i(\cdot)$ represents idiosyncratic noise of the traders, and $\varepsilon(t)$ is a common noise field.

Taking in the Definition 3.8 for the potential $u(\{i\}) = a\nu_i(t) + b\varepsilon(t)$, then $Y_i(t)$ can be considered to be a (time dependent) energy of a "Gibbsian process" with a nearest neighborhood pair potential. For more details, see [Voi03, p. 236].

Although this development is up to now in accordance with our synchronous and local transition operators, it turns out that the wealth process (6.12) itself does not share the locality property. This is due to the fact that the new prices are generated by a mechanism that incorporates global trading information in a way that resembles (6.15).

Summarizing: The market models share to a certain extent the Markov property with restricted memory in space and time for our time dependent Markov random fields, but there is — at least in the models described here — common global information required that violates the standard definitions of Markov random fields. This would even continue to hold if we enlarge the state space in a way to include the T-past of the moving averages into the actual state of the systems by the supplemented variable technique.

BIBLIOGRAPHY

[AH95]   E. Altman and A. Hordijk. Zero-sum Markov games and
           worst-case optimal control of queueing systems. *QUESTA*,
           21:415 447, 1995.

[AK78]   W. J. Arkin and A. I. Krechetov. Markovian controls in prob-
           lems with discrete time. *Verojatnostnye processy i upravlenie*,
           pages 8 41, 1978.

[AKU02]   A. Anagnostopoulos, I. Kontoyiannis, and E. Upfal. Steady
           state analysis of balanced-allocation routing. Preprint Com-
           puter Science Department, Brown University, Providence,
           2002.

[AL72]   W. J. Arkin and W. L. Levin. Convexity of values of vectorial
           integrals, theorems of measurable selection and variational
           problems. *Uspehi Matematicheskih Nauk*, 28(3):165, 1972.

[Alt96]   E. Altman. Nonzero-sum stochastic games in admission, ser-
           vice and routing control in queueing systems. *QUESTA*,
           23:259 279, 1996.

[Alt99]   E. Altman. A Markov game approach for optimal routing
           into a queueing network. In M. Bardi, T. E. S. Raghavan,
           and Parthasarathy T., editors, *Stochastic and Differential
           Games*, volume 4 of *Annals of the International Society of
           Dynamic Games*, pages 359 376. Birkhäuser, Boston, 1999.

[And91]   W. J. Anderson. *Continuous Time Markov Chains*. Springer,
           New York, 1991.

[Ass83]   D. Assaf.   Extreme-point solutions in Markov decision processes. *J. Appl. Probab.*, 20(4):835–842, 1983.

[Ave72]   M. B. Averintsev. Description of Markov random fields by means of Gibbs conditional probabilities. *Theor. Veroyatnost. Primen.*, 17:21–35, 1972. [in Russian].

[Bal89]   E. J. Balder. On compactness of the space of policies in stochastic dynamic programming. *Stochastic Processes and Their Appplications*, 32:141–150, 1989.

[Bat76]   J. Bather. Optimal stationary policies for denumerable Markov chains in continuous-time. *Adv. in Appl. Probab.*, 8:114–158, 1976.

[Bel97]   Yu. K. Belyaev. The continuity theorem and its application to resampling from sums of random variables. *Theory of Stochastic Processes*, 3 (19)(1–2):100–109, 1997.

[Bel00]   Yu. K. Belyaev. Unbiased estimation of accuracy of discrete colored image. *Teor. Ymovirnost. Matem. Statist.*, 63:13–20, 2000. [in Russian].

[Ber87]   D. P. Bertsekas. *Dynamic Programming: Deterministic and Stochastic Models.* Prentice-Hall, Englewood Cliffs, NJ, 1987.

[BGM69]   Yu. K. Belyaev, Yu. I. Gromak, and V. A. Malyshev. On invariant random boolean fields. *Matem. Zametki*, 6(5):555–566, 1969. [in Russian].

[BSdL97]   Yu. K. Belyaev and S. Sjöstedt-de Luna. Resampling from independent heterogeneous random variables with varying mean values. *Theory of Stochastic Processes*, 3 (19)(1–2):121–131, 1997.

[BSdL00]   Yu. K. Belyaev and S. Sjöstedt-de Luna. Weakly converging sequences of random distributions. *Journal of Applied Probability*, 37:807–822, 2000.

[BSL90]   J. C. Bean, R. L. Smith, and J. B. Lasserre. Denumerable
           state non-homogeneous Markov decision processes. *J. Math.
           Anal. and Appl.*, 153(1):64–77, 1990.

[Can84]   L. Cantaluppi. Optimality of piecewise-constant policies in
           semi-Markov decision chains. *SIAM Journal of Control and
           Optimization*, 22:723–739, 1984.

[CCG96]   R. Cavazos-Cadena and E. Gaucherand. Value iteration in
           a class of average controlled Markov chain with unbonded
           costs: Necessary and sufficient conditions for poinwise con-
           vergence. *J. Appl. Probab.*, 33:986–1002, 1996.

[CDK04]   R. K. Chornei, H. Daduna, and P. S. Knopov. Stochastic
           games for distributed players on graphs. *Math. Methods of
           Oper. Res.*, 60(2):279–298, 2004.

[CDK05]   R. K. Chornei, H. Daduna, and P. S. Knopov. Controlled
           Markov fields with finite state space on graphs. *Stochastic
           Models*, 21:847–874, 2005.

[Che90]   M. F. Chen. On three classical problems for Markov chains
           with continuous time parameters. *J. Appl. Probab.*, 28:305–
           320, 1990.

[Cho99]   R. K. Chornei. On stochastic games on the graph. *Kiber-
           netika i Sistemny Analiz*, (5):138–144, 1999. [in Russian].
           English translation in: *Cybernetics and Systems Analysis*,
           35(5), September–October 1999.

[Cho01]   R. K. Chornei. On a problem of control of Markovian
           processes on a graph. *Kibernetika i Sistemny Analiz*, (2):159–
           163, 2001. [in Russian]. English translation in: R. K. Chor-
           ney. A Problem of Control of Markovian Processes on a
           Graph. *Cybernetics and Systems Analysis*, 37(2):271–274,
           March–April 2001.

[Chr92]   G. Christakos. *Random Field Models in Earth Sciences*. Academic Press, San Diego, 1992.

[Chu60]   K. L. Chung. *Markov Chains with Stationary Transition Probabilities*. Springer, Berlin, 1960.

[Čin75]   E. Çinlar. *Introduction to Stochastic Processes*. Prentice-Hall, Inc., Englewood Cliffs, New Jersey, 1975.

[Cre99]   N. A. C. Cressie. *Statistics for Spatial Data*. Wiley, New York, 1999. Revised Edition.

[CY01]   H. Chen and D. D. Yao. *Fundamentals of queueing networks*. Springer, Berlin, 2001.

[Dad97]   H. Daduna. Some results for steady-state and sojourn time distributions in open and closed linear networks of Bernoulli servers with state-dependent service and arrival rates. *Performance Evaluation*, 30:3–18, 1997.

[Dad01]   H. Daduna. Stochastic networks with product form equilibrium. In D. N. Shanbhag and C. R. Rao, editors, *Stochastic Processes: Theory and Methods*, volume 19 of *Handbook of Statistics*, chapter 11, pages 309–364. Elsevier Science, Amsterdam, 2001.

[DD02]   B. Desert and H. Daduna. Discrete time tandem networks of queues: Effects of different regulation schemes for simultaneous events. *Performance Evaluation*, 47:73–104, 2002.

[Der63]   C. Derman. On sequential decisions and Markov chains. *Management Science*, 9:16–24, 1963.

[Der64]   C. Derman. Markovian sequential decision processes. *Procedings Symposia Appl. Math.*, 16(281–289), 1964.

[Der70]   C. Derman. *Finite State Markovian Decision Processes*. Academic Press, New York, London, 1970.

[DF93]   P. A. David and D. Foray. Percolation structures, Markov
           random fields. the economics of edi standards diffusions.
           In Pogorel, editor, *Global telecommunications strategies and
           technological changes*. North-Holland, Amsterdam, 1993.
           First version: Technical report, Center for Economical Policy
           Research, Stanford University, 1992.

[DFD98]   P. A. David, D. Foray, and J.-M. Dalle. Marshallian exter-
           nalities and the emergence and spatial stability of technolog-
           ical enclaves. *Economics of Innovation and New Technology*,
           6(2–3), 1998. First version: Preprint, Center for Economical
           Policy Research, Stanford University, 1996.

[DK87]   R. L. Disney and P. C. Kiessler. *Traffic Processes in Queue-
           ing Networks: A Markov Renewal Approach*. The Johns Hop-
           kins University Press, London, 1987.

[DKC01]   H. Daduna, P. S. Knopov, and R. K. Chornei. Local con-
           trol of interacting Markov processes on graphs with compact
           state space. *Kibernetika i Sistemny Analiz*, (3):62–77, 2001.
           [in Russian]. English translation in: Local Control of Markov-
           ian Processes of Interaction on a Graph with a Compact Set
           of States. *Cybernetics and Systems Analysis*, 37(3):348–360,
           May 2001.

[DKC03]   H. Daduna, P. S. Knopov, and R. K. Chornei. Con-
           trolled semi-Markov fields with graph-structured compact
           state space. *Teor. Ymovirnost. Matem. Statist.*, 69:38–51,
           2003. [in Ukrainian]. English translation in: Controlled semi-
           Markov fields with graph-structured compact state space.
           Theory of Probability and Mathematical Statistics, 69:39–53,
           2004.

[DKT78]   R. L. Dobrushin, V. I. Kryukov, and A. L. Toom, editors. *Lo-
           cally Interacting Systems and Their Application in Biology*,
           volume 653 of *Lecture Notes in Mathematics*. Procedings of

the School-Seminar on Markov Interaction Processes in Biology, Held in Pushino, 1976, Springer-Verlag, Berlin, 1978.

[DO82]   V. K. Demin and A. A. Osadchenko. Decomposition by task solution of controlled Markovian processes. *Kibernetika (Kiev)*, (3):98–103, 1982. [in Russian].

[Dob68]   R. L. Dobrushin. Gibbs random fields for a lattice systems with pair-wise interaction. *Funk. Analiz i ego Prilozh.*, 2(4):31–43, 1968. [in Russian].

[Dob71]   R. L. Dobrushin. Markovian processes with many locally interactive components: existence of limit process and its ergodicity. *Probl. Pered. Inform.*, 7(2):149–161, 1971. [in Russian].

[Don79]   Z. Q. Dong. Continuous time Markov decision programming with average reward criterion — countable state and action space. *Sci. Sinica*, SP(II):131–148, 1979.

[Doo53]   J. L. Doob. *Stochastic Processes*. Wiley, New York, 1953.

[Dos76]   Bharat T. Doshi. Continuous time control of Markov processes on an arbitrary state: average return criterion. *Stochast. Process. and Appl.*, 4(1):55–77, 1976.

[DS78]   R. L. Dobrushin and Ya. G. Sinay, editors. *Multicomponent Random Systems*. Nauka, Moskow, 1978. [in Russian].

[DS80]   R. L. Dobrushin and Ya. G. Sinai, editors. *Multicomponent Random Systems*, volume 6 of *Advances in Probability and Related Topics*. Marcel Dekker, New York, 1980. [Translation of the Russian edition 1978].

[DY75]   E. B. Dynkin and A. A. Yushkevich. *Controlled Markov processes and their applications*. Nauka, Moskow, 1975. [in Russian].

[DY95]   M. A. H. Dempster and J. J. Ye. Impulse control of piece-
            wise deterministic Markov processes. *Ann. Appl. Probab.*,
            5(2):399–423, 1995.

[Dyn63]   E. B. Dynkin. *Markov processes.* Fizmatgiz, Moskow, 1963.
            [in Russian]. English translation in: *Markov processes. Vols.
            1–2.* Die Grundlehren der Math. Wissenschaften 121–122,
            Springer-Verlag, Berlin — Gottingen — Heidelberg, 1965.

[Erm76]   Yu. M. Ermoliev. *Methods of Stochastic Programming.*
            Nauka, Moscow, 1976.

[Fai82]   E. A. Fainberg. Controlled Markovian processes with ar-
            bitrary numerical criteria. *Theor. Veroyatnost. Primen.*,
            27(3):456–473, 1982. [in Russian].

[Fai98]   E. A. Fainberg. Continuous time discounted jump Markov
            decision processes: A discrete-event approach. Preprint,
            1998.

[Fed78]   A. Federgruen. On n-person stochastic games with denumer-
            able state space. *Adv. Appl. Probab.*, 10:452–471, 1978.

[Fel68]   W. Feller. *An Introduction to Probability Theory and Its
            Applications: Vol. 1*, volume 1. 3rd ed., John Wiley and
            Sons, Inc., New York, 3 edition, 1968.

[Fel71]   W. Feller. *An Introduction to Probability Theory and Its
            Applications: Vol. 2*, volume 2. 2rd ed., John Wiley and
            Sons, Inc., New York, 2 edition, 1971.

[Fri73]   E. B. Frid. On stochastic games. *Theor. Veroyatnost. Pri-
            men.*, 18:408–413, 1973. [in Russian].

[Fri83]   V. Frishing. Controlled Markov processes and related semi-
            group of operators. *Bull. Austral. Math. Soc.*, 28(3):441–442,
            1983.

[FV96]   J. Filar and K. Vrieze. *Competitive Markov Decision Processes*. Springer, New York, 1996.

[Geo88]   H.-O. Georgii. *Gibbs Measures and Phase Transitions*. Walter de Gruyter, Berlin, 1988.

[GH92]   A. Gravey and G. Hebuterne. Simultaneity in discrete-time single server queues with Bernoulli inputs. *Performance Evaluation*, 14:123–131, 1992.

[GHL03]   X. Guo and O. Hernandez-Lerma. Continuous-time controlled Markov chains. *Annals of Applied Probability*, 13:363–388, 2003.

[Gim90]   G. L. Gimelfarb. *Statistical Models and Technology for Digital Image Processing of the Earth's Surface*. Thesis for a Doctor's degree (Technical Sciences), Academy of Sciences of Ukrainian SSR, V. M. Glushkov Institute of Cybernetics, Kiev, 1990. [in Russian].

[GKK95]   R. J. Gibbens, F. P. Kelly, and P. B. Key. Dynamic alternative routing. In M. E. Steenstrup, editor, *Routing in Communications Networks*, pages 13–47. Prentice Hall, 1995.

[GL01]   X. P. Guo and K. Liu. A note on optimality conditions for continuous-time Markov decision processes with average cost criterion. *IEEE Trans. Automat. Control*, 46:1984–1989, 2001.

[GN67]   W. J. Gordon and G. F. Newell. Closed queueing networks with exponential servers. *Operations Research*, 15:254–265, 1967.

[Gri73]   G. R. Grimmett. A theorem about random fields. *Bulletin of the London Mathematical Society*, 5:81–84, 1973.

[GS69]   L. G. Gubenko and E. S. Statland. On control of Markovian stagewise processes. In *Teor. Optimal Reshen.*, volume 4, pages 24–39. Institute of Cybernetics, Academy of Sciences of Ukrainian SSR, Kiev, 1969. [in Russian].

[GS72a]   L. G. Gubenko and E. S. Statland. On controlled Markov and semi-Markov models and some concrete problems in optimization of stochastic systems. In *Procedings of the Conference on Controlled Stochastic Systems*, pages 87–119. Kiev, 1972.

[GS72b]   L. G. Gubenko and E. S. Statland. On controlled Markov processes in discrete time. *Teor. Verojatnost. Mat. Stat.*, (7):51–64, 1972. [in Russian]. English translation in: *Theory of Probability and Mathematical Statistics*, 7:47–61, 1975.

[GS72c]   L. G. Gubenko and E. S. Statland. On controlled semi-Markov processes. *Kibernetika*, (2):26–29, 1972. [in Russian].

[GS79]   I. I. Gihman and A. V. Skorohod. *Controlled Stochastic Processes*. Springer, New York, 1979.

[Gub72]   L. G. Gubenko. On multistage stochastic games. *Teor. Verojatnost. Mat. Stat.*, (8):35–49, 1972. [in Russian].

[GZ89]   G. L. Gimelfarb and A. V. Zalesnyi. Models of Markovian random fields in problems by generation and segmenting of texture pictures. In *Means for Intellectualization of Cybernetic Systems*, pages 27–36. Institute of Cybernetics, Academy of Sciences of Ukrainian SSR, Kiev, 1989. [in Russian].

[GZ02a]   X. P. Guo and W. P. Zhu. Denumerable-state continuous-time Markov decision processes with unbounded transition and reward under the discounted criterion. *J. Appl. Probab.*, 39:233–250, 2002.

[GZ02b]   X. P. Guo and W. P. Zhu. Optimality conditions for CT-
          MDP with average cost criterion. In Z. T. Hou, J. A. Filar,
          and A. Y. Chen, editors, *Markov Processes and Controlled
          Markov Chains*, chapter 10. Kluwer, Dordrecht, 2002.

[Hil79]   T. P. Hill. On the existence of good markov strategies. *Trans.
          Amer. Math. Soc.*, 247:157–176, 1979.

[Hin70]   K. Hinderer. Foundations of non-stationary dynamic pro-
          gramming with discrete time parameter. In *Lecture Notes in
          Operations Research and Mathematical Systems*, volume 33.
          Springer-Verlag, Berlin — Heidelberg — New York, 1970.

[HL94]   O. Hernandes-Lerma. *Lectures on Continuous-time Markov
          Control Processes*. Sociedad Matemática Mexicana, México
          City, 1994.

[HMRT01] B. R. Haverkort, R. Marie, G. Rubino, and K. Trivedi. *Per-
          formability Modeling, Technique and Tools*. Wiley, New York,
          2001.

[Hos98]   Mitsuhiro Hoshino. On some continuous time discounted
          Markov decision process. *Nihonkai Math. J.*, 9(1):53–61,
          1998.

[How60]   R. A. Howard. *Dynamic Programming and Markov Pro-
          cesses*. Technology Press and Wiley, New York, 1960.

[How64]   R. A. Howard. Research in semi-Markov decision structures.
          Journal of the Operational Research Society of Japan, 6:163–
          199, 1964.

[HP98]   M. Haviv and M. L. Puterman. Bias optimality in controlled
          queuing systems. *J. Appl. Probab.*, 35:16–150, 1998.

[HS84]   D. P. Heyman and M. J. Sobel. *Stochastic Models in Opera-
          tions Research*, volume 2. McGraw-Hill, New York, 1984.

[Hu96]   Qiying Hu. Continuous time Markov decision processes with
            discounted moment criterion.   *J. Math. Anal. and Appl.*,
            203(1):1–12, 1996.

[Hun83]   J. J. Hunter. *Mathematical Techniques of Applied Probability*,
            volume II: Discrete Time Models: Techniques and Applica-
            tions. Academic Press, New York, 1983.

[HV69]   C. J. Himmelberg and F. S. van Vleck. Some selection theo-
            rems for measurable functions. *Canadian Mathematical Jour-
            nal*, 21:394–199, 1969.

[HW78]   K. M. van Hee and J. Wessels. Markov decision processes
            and strongly exessive functions. *Stochast. Process. and Appl.*,
            8(1):59–76, 1978.

[Ila02]   A. Ilachinski.   *Cellular Automata — A Discrete Universe.*
            World Scientific, Singapore, 2002. Reprint of the first edition
            from 2001.

[Jac63]   J. R. Jackson. Jobshop-like queueing systems. *Management
            Science*, 10:131–142, 1963.

[Jaq75]   Stratton C. Jaquette.   Markov decision processes with a
            new optimality criterion: Continuous time. *Ann. Statist.*,
            3(2):547–553, 1975.

[Jew63a]   W. S. Jewell. Markov-renewal programming. I. Formulation,
            finite return models. *Operations Research*, 11(6):938–948,
            1963.

[Jew63b]   W. S. Jewell. Markov-renewal programming. II. Infinite re-
            turn models, example. *Operations Research*, 11(6):948–971,
            1963.

[Kak71]   P. Kakumanu. Continuously discounted Markov decision
            model with countable state and action space. *Ann. Math.
            Statist.*, 42:919–926, 1971.

[Kak72]   P. Kakumanu. Non-discounted continuous-time Markovian decision process with countable state space. *SIAM J. Contr.*, 10(1):210–220, 1972.

[Kak75]   P. Kakumanu.   Continuous time Markovian decision processes average return criterion. *J. Math. Anal. and Appl.*, 52(1):173–188, 1975.

[Kak78]   P. Kakumanu. Solutions of continuous-time Markovian decision models using infinite linear programming. *Nav. Res. Log. Quart.*, 25(3):431–443, 1978.

[KC98]   P. S. Knopov and R. K. Chornei. Controlled Markovian processes with an aftereffect. *Kibernetika i Sistemny Analiz*, (3):61–70, 1998. [in Russian]. English translation in Knopov P. S., Chornei R. K. Problems of Markov Processes Control with Memory. *Cybernetics and Systems Analysis*, 34(3), May 1998.

[KC00]   M. S. Kaiser and N. Cressie. The construction of multivariate distributions from Markov random fields. *Journal of Multivariate Analysis*, 73:199–220, 2000.

[Kel79]   F. P. Kelly. *Reversibility and Stochastic Networks*. John Wiley and Sons, Chichester — New York — Brisbane — Toronto, 1979.

[Kit87]   M. Yu. Kitaev. Elimination of randomization in semi-Markov decision models with average cost criterion. *Optimization*, 18:439–446, 1987.

[KK87]   H. Kawai and N. Katon. Variance constained Markov decision process. *J. Oper. Res. Soc. Jap.*, 30(1):88–100, 1987.

[KK92]   W. Kwasnicki and H. Kwasnicki. Market, innovation, competition: An evolutionary model of industrial dynamics. *Jour-

nal of Economic Behavior and Organisation, 19:343–368, 1992.

[KKN80]   Yu. M. Kaniovsky, P. S. Knopov, and Z. V. Nekrylova. *Limit theorems for stochastic programming processes*. Naukova Dumka, Kiev, 1980. [in Russian].

[KL95]   Y. A. Korilis and A. A. Lazar. On the existence of equilibria in noncooperative optimal flow control. *Journal of the ACM*, 42:584–613, 1995.

[Kno96]   P. S. Knopov. Markov fields and their applications in economics. *Computational and Applied Mathematics*, 80:33–46, 1996. [in Russian].

[Koe79]   G. J. Koeler. Value convargence in a generalized Markov decision process. *SIAM J. Contr. and Optim.*, 17(2):180–186, 1979.

[KR95]   M. Yu. Kitaev and V. V. Rykov. *Controlled queueing systems*. CRC Press, Boca Raton — New York — London — Tokyo, 1995.

[KRN65]   K. Kuratowski and C. Ryll-Nardzewski. A general theorem on selectors. *Bull. Acad. Polon. Sc.*, 13:397–402, 1965.

[Kur69]   K. Kuratowski. *Topology*, volume II. Academic Press, New York, 1969. (revised edition).

[KV80]   O. Kozlov and N. Vasilyev. Reversible markov chains with local interactions. In R. L. Dobrushin and Y. G. Sinai, editors, *Multicomponent Random Systems*, volume 6 of *Advances in probability and related topics*, pages 451–469. Mracel Dekker, New York, 1980.

[Ky53]   Fan Ky. Minimax theorems. *Proceeding of the National Academy of Science U.S.A.*, 39:42–47, 1953.

[Las94]   J. B. Lasserre. Detectiong optimal and non-optimal actions
            in average-cost Markov decision processes. *J. Appl. Probab.*,
            31(4):979–990, 1994.

[Lef81]   C. Lefèvre. Optimal control of a birth and death epidemic
            process. *Oper. Res.*, 29:971–982, 1981.

[LFT77]   G. de Leve, A. Federgruen, and H. C. Tijms. A general
            Markov decision method. i: Model and techniques. *Adv.
            Appl. Probab.*, 9(2):296–315, 1977.

[Lig85]   T. M. Liggett. *Interacting Particle Systems*, volume 276 of
            Grundlehren der mathematischen Wissenschaften. Springer,
            Berlin, 1985.

[Lig99]   T. M. Liggett. *Stochastic interacting systems: Contact,
            voter, and exclusion processes*. Springer-Verlag, Berlin, 1999.

[LP00]   M. E. Lewis and M. Puterman. A note on bias optimality in
            controlled queueing systems. *J. Appl. Probab.*, 37:300–305,
            2000.

[LP01]   M. E. Lewis and M. Puterman. A probabilistic analysis of
            bias optimality in unichain Markov decision processes. *IEEE
            Trans. Automat. Control*, 46:96–100, 2001.

[LT91]   H.-C. Lai and K. Tanaka. On continuous-time discounted
            stochastic dynamic programming. *Appl. Math. and Optimiz.*,
            23(2):155–169, 1991.

[LVRA94]   F. Luque Vasquez and M. T. Robles Alcaraz. Controlled
            semi-Markov models with discounted unbounded costs. *Bol.
            Soc. Math. Mex.*, 39:51–68, 1994.

[Mar83]   O. N. Martynenko. Finding of optimal control strategy
            for one class of Markovian decision processes. *Kibernetika
            (Kiev)*, (2):117–118, 1983. [in Russian].

[Mil68]   B. L. Miller. Finite state continuous time Markov decision
            processes with infinite planning horizon. *J. Math. Anal.
            Appl.*, 22(3), 1968.

[MM85]   V. A. Malyshev and R. A. Minlos. *Gibbs random fields.*
            Nauka, Moscow, 1985. [in Russian].

[MP70]   A. Maitra and T. Parthasarathy. On stochastic games. *J.
            Optimization Theory and Appl.*, 5:289–300, 1970.

[MS96]   A. P. Maitra and W. D. Sudderth. *Discrete Gambling and
            Stochastic Games*, volume 32 of *Application of Mathematics.*
            Springer-Verlag, New York, 1996.

[MY66]   A. S. Monin and A. M. Yaglom. *Statistical Hydromechanics*,
            volume 1. Nauka, Moskow, 1966. [in Russian].

[MY67]   A. S. Monin and A. M. Yaglom. *Statistical Hydromechanics*,
            volume 2. Nauka, Moskow, 1967. [in Russian].

[Mye84]   R. B. Myerson. Cooperative games with incomplete infor-
            mation. *Internat. J. Game Theory*, 13:69–96, 1984.

[Neu56]   J. von Neumann. Probabilistic logics and the synthesis of
            reliable organisms from unreliable components. In C. E.
            Shannon and J. McCarthy, editors, *Automata Studies*, pages
            43–98. Princeton, 1956.

[NH73]   M. B. Nevelson and R. Z. Hasminskii. *Stochastic Approxi-
            mation and Recursive Estimation.* AMS, Providence, Rhode
            Island, 1973.

[NS99]   A. S. Nowak and K. Szajowski. Non-zero sum stochas-
            tic games. In M. Bardi, T. E. S. Raghavan, and T. Par-
            thasarathy, editors, *Stochastic and Differential Games*, vol-
            ume 4 of *Annals of the Internatioanl Society of Dynamic
            Games*, pages 297–342. Birkhäuser, Boston, 1999.

[PCS86]   G. Pujolle, J. P. Claude, and D. Seret. A discrete queue-
          ing system with a product form solution. In T. Hasegawa,
          H. Takagi, and Y. Takahashi, editors, *Proceedings of the
          IFIP WG 7.3 International Seminar on Computer Network-
          ing and Performance Evaluation*, pages 139–147. Elsevier
          Science Publisher, Amsterdam, 1986.

[Pre74]   C. Preston. *Gibbs States on Countable Sets*. Cambridge
          University Press,London, 1974.

[Put94]   M. L. Puterman. *Markov Decision Processes*. Wiley, New
          York, 1994.

[Ren93]   E. Renshaw. *Modelling Biological Populations in Space and
          Time*, volume 11 of *Cambridge Studies in Mathematical Bi-
          ology*. Cambridge University Press, Cambridge, paperback
          edition edition, 1993.

[RF91]   T. E. S. Raghavan and J. A. Filar. Algoritms for stochas-
          tic games — a survey. *Zeitschrift für Operations Research*,
          35:437–472, 1991.

[Rip81]   B. D. Ripley. *Spatial Statistics*. Wiley, New York, 1981.

[Ros70]   S. M. Ross. Average cost semi-Markov processes. *Journal of
          Applied Probability*, 7:649–656, 1970.

[Roz81]   Yu. A. Rozanov. *Markov Random Fields*. Nauka, Moskow,
          1981. [in Russian].

[Roz82]   Yu. A. Rozanov. *Markov Random Fields*. Springer, New
          York, 1982. [translation of the Russian version of 1981].

[Rue69]   D. Ruelle. *Statistical Mechanics*. W. A. Benjamin Inc., New
          York — Amsterdam, 1969.

[Sch75]   M. Schäl. Conditions for optimality in dynamic programming and for the limit of n-stage optimal policies to be optimal. *Zeitschrift für Wahrscheinlichkeitstheorie und verwandte Gebiete*, 32:179–196, 1975.

[Sen99]   L. I. Sennot. *Stochastic Dynamic Programming and the Control of Queueing Systems*. Wiley, New York, 1999.

[Ser79]   R. F. Serfoso. An equivalence between continuous and discrete time Markov decision processes. *Oper. Res.*, 27(3):616–620, 1979.

[Ser81]   R. F. Serfoso. Optimal control of random walks, birth and death processes, and queues. *Adv. in Appl. Probab.*, 13:61–83, 1981.

[Ser92]   R. F. Serfoso. Reversibility and compound birth death and migration processes. In *Queueing and Related Models*, pages 65–90. Oxford University Press, Oxford, 1992.

[Ser99]   R. F. Serfozo. *Introduction to Stochastic Networks*, volume 44 of *Applications of Mathematics*. Springer, New York, 1999.

[Sha53]   L. S. Shapley. Stochastic games. *Proc. Nat. Acad. Sci. USA*, 39:1095–1100, 1953.

[Sin80]   Ya. G. Sinay. *The Theory of Phase Transitions*. Nauka, Moscow, 1980. [in Russian].

[Sko90]   A. V. Skorohod. *Lections on theory of random processes*. Lybid', Kyiv, 1990. [in Ukrainian].

[SS86]   M. Schäl and W. Sudderth. Stationary policies and Markov policies in Borel dynamic programming. *Probab. Theory and Relat. Fields*, 74(1):91–111, 1986.

[Sta71]   O. N. Stavskaya. Gibbs invariant measure for Markov chains on finite lattices with locally interaction. *Mathematical Collection*, 92(3):402–419, 1971. [in Russian].

[Str66]   R. E. Strauch. Negative dynamic programming. *Ann. Math. Statist.*, 37:871–890, 1966.

[Str05]   D. W. Stroock. *An Introduction to Markov Processes.* Springer, Berlin, 2005.

[Sul75]   W. G. Sullivan. *Markov Processes for Random Fields.* Dublin, 1975.

[SV02]   G. Silverberg and B. Verspagen. A percolation model of innovation in complex technology spaces. Technical Report 2002-5, MERIT-Maastricht Economic Research Institute on Innovation and Technology, Maastricht, 2002. Second draft.

[Tij94]   H. C. Tijms. *Stochastic Models: An Algorithmic Approach.* Wiley, Chichester, 1994.

[VA93]   O. Vega-Amaya. Average optimality in semi-Markov control models on Borel spaces: Unbounded cost and controls. *Bol. Soc. Math. Mex.*, 38:47–60, 1993.

[Vas78]   N. B. Vasilyev. Bernoulli and Markov stationary measures in discrete local interactions. In R. L. Dobrushin, V. I. Kryukov, and A. L. Toom, editors, *Locally Interacting Systems and Their Application in Biology*, volume 653 of *Lecture Notes in Mathematics*, pages 99–112. Procedings of the School-Seminar on Markov Interaction Processes in Biology, Held in Pushino, 1976, Springer-Verlag, Berlin, 1978.

[Ver80]   D. Vermes. On the semigroup theory of stochastic control. *Lect. Notes Contr. and Inf. Sci.*, 25:91–102, 1980.

[Ver81]   D. Vermes. Optimal stochastic control under reliability consraints. *Lect. Notes Contr. and Inf. Sci.*, 36:227–234, 1981.

[VK78]   N. B. Vasilyev and O. K. Kozlov. Reversible Markov chains with local interaction. In *Many-component random systems*, pages 83–100. Nauka, Moscow, 1978. [in Russian].

[Voi03] J. Voit. *The Statistical Mechanics of Financial Markets*. Springer, Berlin, 2 edition, 2003.

[VS64]   O. V. Viskov and A. N. Shiryayev. On controls leading to optimal stationary states. *Trudy Mat. Inst. Steklov*, 71:35–45, 1964. [in Russian]. English translation in: *Selected Translations in Mathematical Statistics and Probability*, 6:1966, 71–83.

[Wak87] K. Wakuta. Arbitrary state semi-Markov decision processes with unbounded rewards. *Optimization*, 18:447–454, 1987.

[Wal88] J. Walrand. *An Introduction to Queueing Networks*. Prentice-Hall, Englewood Cliffs, NJ, 1988.

[Whi86] P. Whittle. *Systems in Stochastic Equilibrium*. Wiley, Chichester, 1986.

[Whi88] D. J. White. Discount-isotone policies for Markov decision processes. *OR Spectrum*, 10(1):13–22, 1988.

[Win95] G. Winkler. *Image Analysis, Random Fields and Dynamic Monte Carlo Methods*, volume 27 of *Applications of Mathematics*. Springer, Heidelberg, 1995.

[Yad80] M. I. Yadrenko. *Spectral Theory of Random Fields*. Vyshcha shkola, Kyiv, 1980. [in Russian].

[Yao95] D. Yao. S-modular games, with queueing applications. *Queueing Systems and Their Applications*, 21:449–475, 1995.

[YF79]   A. A. Yushkevich and E. A. Fainberg. On homogeneous Markov model with continuous time and finite or countable state space. *Theor. Veroyatnost. Primen.*, 24(1):155–160, 1979. [in Russian]. English translation in: On homogeneous Markov model with continuous time and finite or countable state space. *Theory Probab. Appl.*, 24(1):156–161, 1979.

[Yus77]   A. A. Yushkevich. Controlled Markov models with countable state set and continuous time. *Theor. Veroyatnost. Primen.*, 22(2):222–241, 1977. [in Russian].

[Yus80]   A. A. Yushkevich.   Controlled jump Markovian models. *Theor. Veroyatnost. Primen.*, 25(2):247–270, 1980.   [in Russian].

[Yus83]   A. A. Yushkevich.   Continuous time Markov decision processes with interventions. *Stochastics*, 9(4):235–274, 1983.

[Zal91]   A. V. Zalesnyi. *Algorithms for Digital Image Processing Describable by Markovian Random Fields*. Ph. D. thesis (Technical Sciences), Academy of Sciences of Ukrainian SSR, V. M. Glushkov Institute of Cybernetics, Kiev, 1991. [in Russian].

[Zhe91]   S. Zheng. Continuous time Markov decision programming with average reward criterion and unbounded reward rate. *Acta math. appl. sin. Engl. Ser.*, 7(1):6–16, 1991.

Index